中斯友谊的象征

“纪念班达拉奈克国际会议大厦”援建工程技术及纪实

A Symbol of China-Sri Lankan Friendship

The Documentary and Engineering Technology of Bandaranaike Memorial International Conference Hall

中英文对照

Chinese-English Edition

由宝贤等　编著

You Baoxian etc.

中国建筑工业出版社

China Architecture & Building Press

图书在版编目（CIP）数据

中斯友谊的象征“纪念班达拉奈克国际会议大厦”援建工程技术及纪实 中英文对照 / 由宝贤等编著. —北京：中国建筑工业出版社，2013.4
ISBN 978-7-112-15244-5

Ⅰ. ①中… Ⅱ. ①由… Ⅲ. ①会堂－建筑艺术－斯里兰卡－画册②中外关系－对外援助－斯里兰卡－画册 Ⅳ. ①TU242.1-64②D822.235.8-64

中国版本图书馆CIP数据核字(2013)第051460号

责任编辑：石振华 白玉美 姚丹宁
责任校对：姜小莲 关 健
Editors: Shi Zhenhua Bai Yumei Yao Danning
Proofreaders: Jiang Xiaolian Guan Jian

中斯友谊的象征
“纪念班达拉奈克国际会议大厦”援建工程技术及纪实

中英文对照

由宝贤等 编著

A Symbol of China-Sri Lankan Friendship
The Documentary and Engineering
Technology of Bandaranaike Memorial International Conference Hall

Chinese-English Edition
You Baoxian etc.

*

中国建筑工业出版社 出版、发行（北京西郊百万庄）
各地新华书店、建筑书店经销
北京美光设计制版有限公司制版
北京方嘉彩色印刷有限责任公司印刷

*

开本：880×1230毫米 1/16 印张：10 3/4 字数：335千字
2013年4月第一版 2015年1月第二次印刷
定价：150.00元

ISBN 978-7-112-15244-5
(23324)

谨以此书庆贺
“纪念班达拉奈克国际会议大厦”
援建工程竣工40周年

Dedicate this book to celebrate
Bandaranaike Memorial International Conference Hall
40th Anniversary of the completion of the project

序言1

中国建筑设计研究院的前身“中央直属设计公司”，系1952年由中共中央办公厅直属修建办事处设计室、国务院中南海建筑公司、中央军委民用航空局设计处、中国建筑工程设计公司等11个中央在京直属单位的建筑设计机构组建而成。我院承担的援外设计任务于成立之初开始，迄今已有60余年，在全世界56个国家和地区都留下了足迹，其中不乏多项建筑设计精品。如20世纪50年代的蒙古乌兰巴托跨线桥、越南机械修配厂；20世纪60年代的斯里兰卡“纪念班达拉奈克国际会议大厦”；20世纪70年代的巴基斯坦综合体育设施；20世纪80年代的也门总统官邸；20世纪90年代的乌干达国家体育场；21世纪初的援非盟会议中心项目等。60多年来，我院见证了新中国援外工作的历程，在受援国的大地上留下了一个个精彩的传世作品，也在受援国人民的心中留下了中国人民的真挚友谊，更为中国维护世界和平与发展，争取人类进步事业做出了贡献。

在我院诸多援外设计作品之中，斯里兰卡“纪念班达拉奈克国际会议大厦”绝对可以称得上是一个可圈可点、经得起历史检验的代表性项目，赢得了国内外各方的好评，至今仍被誉为我国援外工程的典范。40多年过去了，她挺拔的身姿、端庄的外表、靓丽的形象给人以难以忘怀的印象，仍是斯里兰卡首都科伦坡的标志性建筑。1974年，“纪念班达拉奈克国际会议大厦”刚落成，第五次不结盟国家首脑会议即在此召开，立刻成为进行国际对话和讨论的中心。各国与会代表对大会堂的美观和功能都非常满意，之后还召开过多项国际会议，都得到了各个国家的赞许。中共中央办公厅和国务院办公厅基于上述反映，当时还通报表扬了建筑设计团队。“纪念班达拉奈克国际会议大厦”1978年获得国家第一届科技大会设计奖，2009年获得中国建筑学会建国60周年建筑创作大奖，2010年在全国经援工作60周年大会上，与会领导仍对其予以高度评价，“无论是设计上，还是施工质量上，‘纪念班达拉奈克国际会议大厦’迄今为止是最好的。”这个成为中国援外工作典范工程的大厦，至今仍受到斯里兰卡人民的爱戴。她是中斯两国人民友谊的见证，为两国人民的友好交流留下了珍贵的、美好的回忆。

中国建筑设计研究院的老一辈领导和技术专家为祖国援外事业做出了历史性的、奠基性的突出贡献，是青年一代的学习榜样与楷模。新一代中国建筑设计研究院人将以“传承中华文化、打造中国设计、促进科技进步、引领行业发展”为使命，继承并发扬老一辈求实严谨、敬业奉献的光荣传统，锐意创新、团结共进，为祖国建设事业和外交事业再立新功！

中国建筑设计研究院　院长
修龙
二〇一三年二月二十八日

Foreword 1

The predecessor of the China Architecture Design & Research Group (CAG)—the Central Design Company, was founded in 1952 from 11 national architectural design institutions in Beijing, including the design studio of General Office of the Communist Party Central Committee of China; Zhongnanhai Construction Company of the State Council; the Civil Aviation Administration Design Office of the Central Military Commission; China Construction Engineering Design Company, etc. It has been 60 years since CAG undertook design works of foreign aid projects when we first established. We have left our traces in 56 countries and regions around the world. A number of designs are fine-arts, including Mongolia – Ulaanbaatar Overpass and Vietnam Machine Repair Shop in 1950s, Sri Lanka Bandaranaike Memorial International Conference Hall in 1960s, Sports Complex Islamabad Pakistan in 1970s, the President House of Yemen in 1980s, Uganda National Stadium in 1990s, and the design supervision of the aid project of African Union Conference Center in the early 21st century. We have witnessed the course of the new China's foreign aid works in the past 60 years, leaving a number of long lasting splendid works in recipient countries, where people cherish the sincere friendship with the Chinese. The efforts from China in turn contribute to the maintenance of the world peace and development along with the mankind progress.

Among the numerous foreign aid works, the Sri Lanka Bandaranaike Memorial International Conference Hall (the Hall) is definitely a remarkable epic of project even till now, which has won multiple praises domestically and internationally. Over four decades, its dignified appearance and elegant image gives unforgettable impression, which is still a landmark in Colombo, the capital of Sri Lanka.

The Fifth Summit of Non-Aligned Countries was held here right after the Hall was built in 1974, that immediately made it the center of international dialogues and discussions. The delegates of the participant countries were very pleased with the beauty and functions of the Hall. It continues gaining applauds from delegates participating international conferences held in the Hall afterwards. Our design team was thus celebrated by the General Office of the Communist Party of China and the General Office of the State Council of China. The Hall was rewarded with the Design Award from the 1st National Science and Technology Conference in 1978, and the Architectural Award of Creativity at the 60th Anniversary of the Founding of the PRC from the Architectural Society of China in 2009. During the conference of the 60th Anniversary of National Foreign Aid in 2010, the Hall was still highly evaluated. "The Hall so far is one of the bests, in terms of design and the quality of works." said the speaking leader. This model project for China's Foreign Aid is still loved by the Sri Lankan people. It witnessed the friendship between peoples of the two countries, and has left precious memories of friendly exchanges between them.

The senior generation of leaders and experts of CAG have made historic, groundbreaking outstanding contributions to the foreign aid course of the country, who has set the model for the younger generation to follow. The new generation of CAG will take "To heritage the tradition of Chinese culture, To create China Design, To promote the progress of science and technology, and To lead the development of the industry" as their mission, and will inherit and carry forward the senior generation's tradition of "rigorous and truth-seeking; commitment and dedication; keen on innovation", and will make their new contributions to the construction and diplomatic courses of China.

Xiu Long
President

China Architecture Design & Research Group (CAG)

2013-02-28

序言2

（“班厦”管委会负责人班杜拉先生亲笔用僧伽罗文为本书书写的序言）

චීන ශ්‍රී ලංකා සබඳතා සහ බණ්ඩාරනායක
අනුස්මරණ ජාත්‍යන්තර සම්මන්ත්‍රණ ශාලාව.

නූතන චීන - ශ්‍රී ලංකා සබඳතා 1950 වර්ෂයේ ජනවාරි මාසයේදී ඇරඹිනි. චීන - ශ්‍රී ලංකා සබඳතා අතර කැපී පෙනෙන සංසිද්ධියක් ලෙස 1952 දී දෙරට අතර අත්සන් කළ රබර් - සහල් ගිවිසුම හැඳින්විය හැකිය. 1957 පෙබරවාරි මාසයේදී දෙරට අතර රාජ්‍යතාන්ත්‍රික සබඳතා ඇතිවිය. ප්‍රවීණ චීන අගමැතිවරයෙකු වූ චූ එන් ලායි මහතා 1957 දී ශ්‍රී ලංකාවේ සංචාරයක යෙදුනි. ශ්‍රී ලංකාවේ රාජ්‍ය නායකයෙකුගේ ප්‍රථම චීන සංචාරය අගමැති සිරිමාවෝ බණ්ඩාරනායක මැතිනිය විසින් 1961 දී සංවිධානය කරන ලදී. මෙම ඓතිහාසික රාජ්‍යතාන්ත්‍රික චාරිකා දෙරට අතර උණුසුම් මිත්‍රත්වයක් අඛණ්ඩව පැවැත්ම සහතික කරන ලදී.

වර්ෂ 1973 දී තෑගි පරිත්‍යාගයක් ලෙස චීනයෙන් ශ්‍රී ලංකාවට පිරිනැමුණු ඓතිහාසික චීන - ශ්‍රී ලංකා මිත්‍රත්වයේ ජීවමාන සංකේතය මෙම සම්මන්ත්‍රණ ශාලාවයි. BMICH යන සංක්ෂිප්තයෙන් ප්‍රචලිත මෙම සමූහ හා ප්‍රධාන සංකීර්ණය චීනයේ ශිල්පකර්මාන්තයේ අභිමානවත් දෙරටක් අතර පවතින ගැඹුරු මිත්‍රත්වය සංකේතවත් කරයි. ගතවූ සිව් දශකය පුරා ශ්‍රී ලංකාවේ පැවැත්වූ දේශීය සහ ජාත්‍යන්තර බොහෝ සම්මන්ත්‍රණ හා ප්‍රදර්ශන පිළිබඳව ආපසු හැරී බැලූවොත් බණ්ඩාරනායක අනුස්මරණ ජාත්‍යන්තර සම්මන්ත්‍රණ ශාලා පරිශ්‍රය (BMICH පරිශ්‍රය) නොමැතිව අප එවැනි හැකිව සිදු අපහසුතාවයක් ගේ වැටහීමකට අවබෝධ වේ. මෙම පොත මගින් මහජන චීන සමූහාණ්ඩුවේ පරිත්‍යාගයක් ලෙස අපට ලැබුණ මෙම සමූහ සංකීර්ණය පිළිබඳව අපට ඇතිවන අභිමානය, චීනය පිළිබඳව අපගේ කෘතඥතාවය දැක්වීමට අපි කැමැත්තෙමු.

චීන - ශ්‍රී ලංකා මිත්‍රත්වය සැමදා දිනේවා.

序言2

（“班厦”管委会负责人班杜拉先生亲笔用僧伽罗文为本书写的序言的中文译文）

中国和斯里兰卡的关系以及“纪念班达拉奈克国际会议大厦”

当代的中国和斯里兰卡的关系，正式开始于1950年1月，即两国间签署的第一个双边协定——橡胶－大米协定，而两国的外交关系则起始于1957年2月。中国伟大的领导人，周恩来总理，于1957年访问斯里兰卡。斯里兰卡西丽玛沃·班达拉奈克总理1961年访问中国，成为了斯里兰卡历史上第一个到中国访问的国家领导人。

纪念班达拉奈克国际会议大厦是1973年中国无偿赠送给斯里兰卡的礼物。这座宏伟的会展中心，简称为BMICH（“班厦”），骄傲地屹立在斯里兰卡首都的中心地带，是中斯友谊的象征。这座宏伟的建筑从其诞生之日起就是一个繁忙的场所，在许多国家眼里她就是斯里兰卡的象征，因为斯里兰卡的许多国际性活动，都是在这里举行的。

“班厦”，斯中两国建筑风格相结合的成功范例，吸引着所有到此地的参观者。在当今世界中，对任何国家来说拥有一座设施完备的会展中心都是一个加分项。回顾过去的四十年，如果没有“班厦”的存在，那些为斯里兰卡带来巨大荣誉和声望的重大会议和展览，就不可能成功举办。“班厦”以她那令人骄傲的外观，美丽的环境，宽敞的空间，现代化的设施以及赢得的声望，在所有类似设施建筑中占据了最显著的地位。因此，我们感谢中国人民慷慨地赠送给斯里兰卡如此恢宏的礼物。

Foreword 2

(English translation of Preface in Sinhala by Mr. Bandhula Ekanayake, Chief of BMICH)

China-Sri Lanka Relation and the Bandaranaike Memorial International Conference Hall

Modern ties between China and Sri Lanka officially began in January 1950. The first bilateral agreement signed between China and Sri Lanka was the Rubber-Rice Pact. Diplomatic relation, between China & Sri Lanka, came in to the existence in February 1957. A grate Chinese leader former premier, Zhou Enlai visited Sri Lanka in 1957. Prime Minister of Sri Lanka Madam Sirimavo Bandaranaide's visit to China in 1961 marked the first ever visit of a Sri Lankan state leader to China. There historic visits endorsed the continuation of a warm friendship between the two nations.

Bandaranaike Memorial International Conference Hall came to the existence in the form of an outright gift from China to Sri Lanka in the year 1973. This grand convention & Exhibition Centre, widely known as the BMICH, standing proudly in the centre of the capital city of Sri Lanka, symbolizes the profundity of the historic relation maintained by the two countries. This massive creation, which has been highly productive since inception, instrumented also to carry the Sri Lanka's name among the other countries, as it was the venue for many international events held in Sri Lanka.

The BMICH, an example to China-Sri Lanka architectural combination, is definitly an attraction for any visitor passes by the edifice. In the modern world it is a plus factor for a county to own a meeting hall, complete with adequately provided facilities. When looking back of the major and important Conventions & Exhibitions held during the past four decals in Sri Lanka, which brought the honor and fame to the country, wouldn't have been successful if the BMICH was not existed. The proud external appearance, aesthetic surroundings, spaciousness, adequately provided moern facilities, added by the earned reputation, are the factors that leave the BMICH in the most distinct position among similar creations, if any, to come in the time to come. Therefore, it induces us to be grateful to the people of China for been generous to offer this magnificent gift to Sri Lanka.

目录

CONTENTS

“班厦”立面
Façade of BMICH

综述

Introduction

1972年6月28日毛泽东主席在中南海会见斯里兰卡西丽玛沃・班达拉奈克总理
Chairman Mao Zedong met Sri Lankan Prime Minister Sirimavo Bandaranaike in Zhongnanhai on June 28, 1972

新中国成立的时期，也是亚非拉地区被压迫民族和人民争取民族独立和解放革命运动风起云涌的时期，世界上一大批殖民地、半殖民地国家纷纷独立，宣告了帝国主义殖民体系的瓦解。毛泽东主席多次代表中国政府对各国人民的反帝反殖斗争给予坚决支持和高度评价。中国对这些国家的经济援助，表达了对通过民族斗争捍卫民族独立和国家主权的支持。

“纪念班达拉奈克国际会议大厦”（以下简称“班厦”或BMICH）是斯里兰卡政府和人民为纪念斯里兰卡已故的所罗门・韦斯特・里奇韦・迪亚斯・班达拉奈克总理，由我国无偿援建的，位于斯里兰卡首都科伦坡市中心地段。

1965年3月，由总理班达拉奈克夫人（后以“总理班夫人”做亲切称呼）亲自主持了奠基典礼。但由于斯方原因，工程搁置了；1970年11月24日，“班厦”正式动工。1973 年4月竣工，当年5 月17日举行揭幕典礼。

The period during the founding of the People's Republic of China, oppressed nations and people in Asia, Africa and Latin America were struggling for national independence and liberation, and a large number of colonial and semi-colonial countries declared their independence and the collapse of the colonial system of imperialism. Representing the Chinese government, Chairman Mao Zedong gave strong support and appreciated the struggles against imperialism and colonialism by people of these countries and regions. China's economic assistance to these countries demonstrated its support in safeguarding national independence and sovereignty.

Located in downtown Colombo, the capital of Sri Lanka, Bandaranaike Memorial International Conference Hall (BMICH) was an outright gift from the government and people of the People's Republic of China to the government and people of Sri Lanka to commemorate the late Prime Minister Solomon West Ridgeway Dias Bandaranaike.

In March 1965, Prime Minister Sirimavo Bandaranaike personally hosted the BMICH foundation laying ceremony. However, the project was suspended on Sri Lanka's side, and on November 24, 1970, construction on BMICH formally commenced. The project was completed in April 1973, and on May 17 of the same year, an opening ceremony was held to unveil the project.

“纪念班达拉奈克国际会议大厦”总占地面积13公顷，总建筑面积32540平方米，有主体建筑及附属建筑、停车场、喷水池等。

“班厦”是中国援外项目中第一个大型文化建筑项目，同时也是斯里兰卡未来举行重要活动和召开国际会议的场所。

1976年8月，第五次不结盟国家首脑会议在大厦举行，有120多个国家的元首、政府首脑、外长参加。会后，斯里兰卡总理班夫人高兴地对中国大使说：“这次科伦坡会议盛况空前，与会国家最多，开得很成功。要是没有这座大厦，我的本领再大也不敢请他们来啊！各国代表对大厦建筑和内部设备都赞不绝口，说比历次首脑会议的会堂都好。因为会议开得好，就使我更加怀念周恩来总理。”

1977年4月17—22日，总理班夫人为减轻自己怀念的痛苦，特邀邓颖超登上了美丽的岛国斯里兰卡，专门在“班厦”为邓颖超举行盛大宴会。邓颖超不仅看到了周恩来总理一直关心的这座大厦的落成，也看到了周恩来总理当年在佩拉德尼亚热带植物园种下的那棵紫薇，正开着繁茂的花朵。

1983年5月30日，“班厦”建成十周年，斯里兰卡政府举行了隆重的庆祝活动，城乡建设环境保护部部长李锡铭率中国政府代表团应邀参加。

1990年12月18日李鹏总理、2001年5月19日朱镕基总理、2005年4月9日温家宝总理、2012年9月15日吴邦国委员长等国家领导人都曾亲临“班厦”。“班厦”现在还是斯里兰卡首都科伦坡地标性建筑和主要旅游景点之一。

Bandaranaike Memorial International Conference Hall occupies a total land area of 13 hectares, and its floor area is 32,540 square meters. The project includes a main building, annexes, garage and fountain.

As the first large-scale cultural architectural building constructed by China for foreign countries, BMICH is also a venue for Sri Lanka to hold future significant events and international conferences.

In August 1976, the Fifth Summit Conference of the Non-Aligned Nations was held at Bandaranaike Memorial International Conference Hall in Colombo, and more than 120 heads of State and Foreign Affairs ministers attended this conference. After the event, Prime Minister Sirimavo Bandaranaike was pleased told the Chinese ambassador, "This conference was unprecedented, grand and successful with the largest number of participating countries; but if it wasn’t for the convention center, even with our skill and ability, how could we dare to invite these participants! The delegates from various countries highly appreciate the convention center building and its interior equipment, saying it is better than any of the assembly halls of all of the previous summits. The more successful the summit is, the more we miss Premier Zhou Enlai".

From April 17 to 22, 1977, Prime Minister Sirimavo Bandaranaike reminisced about the memories she had with Premier Zhou Enlai, and invited Deng Yingchao, wife of Premier Zhou Enlai to Sri Lanka, the beautiful island country, and hosted a grand and special banquet for Deng at BMICH. Deng Yingchao not only witnessed the establishment of BMICH, which had been Premier Zhou Enlai’s constant concern, but also saw the crape myrtle planted by Premier Zhou Enlai himself in the Peradeniya Botanical Gardens which was in full bloom.

On May 30, 1983, the 10th anniversary of the establishment of BMICH, the Sri Lankan government held a grand celebration, and a delegation from the Chinese government headed by Li Ximing, Minister of Urban and Rural Construction and Environmental Protection was invited to attend the event.

Later, other leaders of the Chinese government, including Premier Li Peng, Premier Zhu Rongji, Premier Wen Jiabao, and National People’s Congress Chairman Wu Bangguo visited BMICH on December 18, 1990, May 19, 2001, April 9, 2005 and September 15, 2012 respectively. For many decades, BMICH has become a landmark building and the premier tourist attraction in Colombo, capital of Sri Lanka.

1. 邓颖超在斯里兰卡受到欢迎
Deng Yingchao was warmly welcomed in Sri Lanka

2. 1977年4月17日，斯里兰卡总理班夫人到机场迎接邓颖超
On April 17, 1977, Prime Minister Sirimavo Bandaranaike welcomed Deng Yingchao at the airport

在“班厦”两侧休息厅摆放毛泽东、周恩来、邓小平的雕像，同时在礼仪大厅班达拉奈克总理雕像旁摆放总理班达拉奈克夫人的雕像，充分体现了斯里兰卡政府对中斯友谊的重视，寄托了对两国友谊奠基者的深切缅怀，也将教育两国年轻一代，牢记老一辈领导人为中斯友谊所做的贡献，继承前辈遗愿，坚定地将中斯友好的火炬世代传递下去。

On both sides of Lounge Hall in BMICH stand the busts of Mao Zedong, Zhou Enlai and Deng Xiaoping, and in Etiquette Hall are those of Mr. and Mrs. Bandaranaike, which fully demonstrate the importance attached by the Sri Lankan government on the China-Sri Lanka friendship. The busts present our deep memory of the founders of the friendship between China and Sri Lanka, which will also teach the younger generation of the two countries, and firmly remember the contributions made of the leaders of the previous generations, and inherit their wills to continue the friendship between China and Sri Lanka in future generations.

1. 2005年4月9日，中国国务院总理温家宝在“班厦”与斯里兰卡总理拉贾帕克萨共同出席周恩来雕像揭幕仪式
On April 9, 2005, Chinese Premier Wen Jiabao and Sri Lankan Prime Minister Mahinda Rajapakse attended the unveiling ceremony of Premier Zhou Enlai's bust in BMICH
2. 礼仪大厅班达拉奈克夫妇雕像
The busts of Mr. and Mrs. Bandaranaike at Etiquette Hall in BMICH

“班厦”在国内曾获奖项

- 1974 年中共中央办公厅及国务院办公厅联合通报表扬。
- 1978 年获第一届全国科技大会设计奖。
- 2009 年获中国建筑学会建国60周年建筑创作大奖。
- 2010 年，在全国纪念经援工作60周年大会上，领导讲话说道：“无论是设计上，还是施工质量上，‘班厦’迄今为止是最好的。”

书中通过让人倍感亲切的黑白图片、赏心悦目的彩色图片、一目了然的“班厦”模型和技术图纸、朴实生动的语言，图文并茂地让读者不但了解了象征着中斯友谊的“班厦”建设的全部历史，还详细记载了“班厦”设计的许多经典技术，也可谓是一部建筑技术知识的宝库。此书既适合中外大众读者阅读，也适合中外技术人员阅读，更具有相关技术人员学习借鉴的价值。

BMICH awards in China

- 1974: The General Office of the CPC Central Committee and the General Office of the State Council of China circulated a notice of commendation of BMICH.
- 1978: The Design Award of the First National Science and Technology Conference.
- 2009: Architectural Creation Award at the 60th Anniversary of the People’s Republic of China organized by the Architectural Society of China.
- 2010: On the occasion of the 60th Anniversary of Celebrating China’s Assistance to Foreign Countries, the leader’s speech said, "In terms of design aral quality of works, BMICH is by far the best".

Through photographs, models, technical drawings and simple yet vivid language, this book recalls the entire story of constructing BMICH, which is a symbol of China-Sri Lanka friendship, and contains many classic technologies for designing BMICH, making it a treasure trove of architectural knowledge. For general and professional readers from China and abroad, this book is a valuable reference.

位于休息大厅的毛泽东（右）与周恩来（左）塑像
The busts of Mao Zedong (right) and Zhou Enlai (left) at Lounge Hall in BMICH

第一章
中斯友谊的象征
Chapter One A Symbol of China-Sri Lankan Friendship

深厚情谊
Profound Friendship

一　米胶协定

1952年12月18日，锡兰[①]政府冲破西方国家对中国的经济封锁，与中国政府在两国还未建交的情况下，在北京签署了《关于大米和橡胶的贸易协定》（简称米胶协定）。缓解了锡兰缺少大米、中国缺少橡胶的困难，那时两国可谓是兄弟患难之交。

二　两国建交

1957年1月31日，周恩来总理、贺龙副总理和随行人员应锡兰政府邀请，到达科伦坡进行友好访问，这是一次名垂青史的访问。

两国总理经过会谈，确认了以和平共处五项原则和万隆会议十项原则作为指导两国关系的准则，并在这一准则指导下，正式建立了外交关系。

1957年2月2日下午，周总理访问康提时，在植物园（佩拉德尼亚热带植物园）种植了一棵紫薇友谊之树，至今茁壮成长，树前有用僧伽罗、泰米尔、英文3种文字书写的牌子。

(I) The Rubber-Rice Pact

On December 18, 1952, the Sri Lankan[①] government broke through the economic blockade imposed by western countries to sign the Rubber-Rice Pact in Beijing with the Chinese government even though the two countries had yet to establish foreign relations. This agreement eased the difficulties the two countries had when Sri Lanka was in short supply of rice, while China was in demand of rubber, making the two countries friends in need.

(II) Establishing Diplomatic Relation

On January 31, 1957, upon invitation of the Sri Lankan government, a Chinese delegation headed by Premier Zhou Enlai and Vice Premier He Long paid a friendly visit to Colombo.

Through an amiable negotiation between the prime ministers of the two countries, China and Sri Lanka reached consensus on having the Five Principles of Peaceful Coexistence and the Ten Principles put forward in the Declaration on the Promotion of World Peace and Cooperation adopted by the Bandung Conference as the guiding principle for relations between the two countries, under which China and Sri Lanka established diplomatic relations.

On February 2, 1957, during his visit to Kandy, Premier Zhou Enlai planted a crape myrtle tree in the Royal Botanical Gardens, Peradeniya. This tree of friendship continues to grow vigorously, and on it is a sign board in Sinhala, Tamil and English.

① 锡兰，是由古代阿拉伯航海家所称的“塞伦底伯”，意为“宝岛”音译演变而来的。我国宋代译为“细兰”；明代译为“锡兰”。在中国古书中，曾以僧伽罗国、师子国、狮子国作为称呼。师子国是我国对其的尊称。僧伽罗国由梵语Simhala一词的译音演变而来；其词义为狮子，狮子国是对僧伽罗国的意译。兰卡，早期音译为“楞伽”，在僧伽罗语中意为“乐土”或“光明富庶的土地”之意，是古称，“斯里”为尊称，无词义。1948年2月4日，斯里兰卡人民经过不屈不挠的斗争，彻底地摆脱了殖民统治，获得了民族的独立，并定国名为锡兰。1972年5月22日，改名为斯里兰卡共和国，1978年8月16日改为斯里兰卡民主社会主义共和国，仍是英联邦成员国。书中除历史部分，自正式建交以后不依其国家的历史时期名称改变而改变，统称：斯里兰卡。

1. 周恩来总理和班达拉奈克总理在一起。左一为班达拉奈克总理
Premier Zhou Enlai with Prime Minister Bandaranaike.Prime Minister Bandaranaike (first left)

2. 周恩来总理和贺龙副总理一行1957年1月31日到达科伦坡时受到盛大欢迎。图为锡兰总理班达拉奈克（左）到机场欢迎周恩来总理（右）
On January 31, 1957, a Chinese delegation headed by Premier Zhou Enlai and Vice Premier He Long was warmly welcomed in Colombo, Sri Lanka.Prime Minister Bandaranaike of Ceylon (left) welcomed Premier Zhou Enlai (right) at the airport

科伦坡独立广场
Independence Square in Colombo

1957年2月4日下午，周恩来总理应邀在首都科伦坡独立广场举行的锡兰独立九周年露天庆祝大会上第一个讲了话。周总理在讲话时，下起了大雨，周总理冒雨演讲，感动了所有在场的人们。

三 独立广场

独立广场是斯里兰卡1948年2月4日独立仪式举行的场所。广场中央的独立纪念堂模仿康提王朝时期皇室接见朝觐者的大厅而建。纪念堂的梁柱上刻有大象、狮子和描述斯里兰卡佛教史的图案等，四周有60个石雕狮子，体现了康提文化的风格。广场中央的地下修建了101个房间，纪念堂的四个角有通往地下的通道。在纪念堂的北面的塑像是斯里兰卡开国总理D·S·森纳那亚克。

《米胶协定》和两国建交、周总理在独立广场冒雨讲话，与斯里兰卡总理班达拉奈克夫妇结下了深厚情谊，成为两国友好合作关系史上的一段佳话。

On February 4, 1957, Premier Zhou Enlai was invited to celebrate the 9th anniversary of Independence Day of Sri Lanka in Independence Square in Colombo and he delivered a speech in the rain, which moved the audience.

(III) Independence Square

Independence Square was the venue for Sri Lanka to hold its independence celebrations on February 4, 1948. Independence Memorial Hall in the center of the square was built based on the hall in which the royal family received pilgrims during the Kandy Dynasty. On the Memorial Hall's beams and columns are carvings of elephants, lions and patterns indicating the Buddhist history of Sri Lanka. Sixty stone lions surround the Memorial Hall, presenting the style of Kandy culture. Underneath the square are 101 rooms that are accessible by passages at the four corners of the Memorial Hall. The statue of Sri Lanka's first Prime Minister, D.S. Senanayake, stands on the north side of the Memorial Hall.

With the Rubber-Rice Pact, diplomatic relations between China and Sri Lanka, the speech in rain delivered by Premier Zhou Enlai, as well as the profound friendship between China and Prime Minister Bandaranaike and his wife, all became a touching story of the friendly cooperation between China and Sri Lanka.

① Ceylon is derived from the word "serendib" named by ancient Arabic sailors, which means "island". It was translated as "xi lan" in the Chinese Song Dynasty, and xi Lan in the Ming Dynasty. In many ancient Chinese books, Ceylon was also called "Sinhalese Country" and "Lion Country". The latter is a respectful name for Ceylon.The former is derived from "Simhala" from sanskrit, which means "Lion". "Lanks" in Sinhalese means "paradise" or "rich and bright land", and "sri" is a respectful name. on February 4, 1948, through persevering struggies against the colonialism, the Sri Lankan people won their national independence, in May 22, 1972, the country was renamed the Republic of Sri Lanka. And on August16, 1978, it was renamed the Democratic Socialist Republic of Sri Lanka, and is still a member of the Commonwealth. In this book, apart from the historical introduction, the country is called Sri Lanka since the establishment of diplomatic relations between China and Sri Lanka.

项目背景
Background of the Project

1. 1964年2月，宋庆龄和周恩来、陈毅访问斯里兰卡
In February 1964, Song Qingling, Zhou Enlai and Chen Yi visited Sri Lanka

2. 1964年2月周恩来总理出访斯里兰卡与总理班夫人亲切交谈
Friendly talks between Premier Zhou Enlai and Prime Minister Sirimavo Bandaranaike in February 1964

3. 斯里兰卡总理班夫人赠周恩来总理的木包银装饰象礼品
A present of silver elephant ornament to Premier Zhou Enlai by Prime Minister Sirimavo Bandaranaike

1964年2月宋庆龄副主席、周恩来总理等国家领导人，为进一步增进新中国同亚非国家的友好合作关系，在出访斯里兰卡（当时称锡兰）期间，中斯两国政府商定，由中国帮助斯里兰卡在科伦坡建造一幢以纪念为国家独立而故去的班达拉奈克总理，并象征中锡两国伟大的友谊的国际会议大厦，当时就命名为"纪念班达拉奈克国际会议大厦"。这一建筑已成为中斯两国人民友好情谊的永久象征。

班达拉奈克总理坚定反对帝国主义和殖民主义，捍卫民族独立的斗争，并对发展斯中两国间的友好关系做出了积极的贡献。

这次访问，斯里兰卡总理班夫人还送给周恩来总理一件木包银装饰象礼品。这件木雕镶银嵌宝石象礼品取材于"象节"。大象驮着的是万人敬仰的斯里兰卡国宝——佛牙塔。木包银装饰象长32厘米，高33厘米。寄托了斯里兰卡总理班夫人对周恩来总理的美好祝愿。

In February 1964, Chinese Vice Chairman Song Qingling and Premier Zhou Enlai visited Sri Lanka to further consolidate friendly cooperation between New China and Asian and African countries. During the visit to Sri Lanka, the two governments agreed to build an international conference building in Colombo to commemorate the late Prime Minister Bandaranaike who fought for the independence of Sri Lanka and to symbolize the great friendship between China and Sri Lanka. The building was named "Bandaranaike Memorial International Conference Hall", and has become a permanent symbol of the friendship between Chinese and Sri Lankan people.

Prime Minister Bandaranaike steadfastly fought against imperialism and colonialism, constantly safeguarded national independence, and made great contributions to develop the China-Sri Lanka friendship.

During this visit, Prime Minister Sirimavo Bandaranaike gave a gift of silver elephant ornament to Premier Zhou Enlai. This present, sourced from the "Elephant Festival", is 32cm long and 33cm high and on its back is the pagoda of the Temple of the Tooth, a national treasure of Sri Lanka, demonstrating the best wishes of Prime Minister Sirimavo Bandaranaike to Premier Zhou Enlai.

“班厦”开工

Commencement of Construction of BMICH

1970年11月24日，锡兰政府为这一象征中锡两国友谊的工程“纪念班达拉奈克国际会议大厦”举行了盛大的开工典礼。锡兰总督威廉·高伯拉瓦和总理班达拉奈克夫人出席了仪式。威廉·高伯拉瓦总督在检阅锡兰陆军仪仗队后，发表了讲话，他对中国政府帮助锡兰建设这座大厦表示感谢，并指出：“这座大厦的建设，将是锡、中两国人民之间友好的象征。”锡兰总理班夫人在讲话中请中国大使向中国人民，特别是向周恩来总理转达锡兰政府诚挚的感谢。中国大使马子卿也讲了话，他赞扬了已故总理班达拉奈克在坚持反对帝国主义和殖民主义、捍卫民族独立斗争和对中、锡两国之间友好关系的发展做出的贡献。他说：“经过我们两国的共同合作，这座大厦的建设将成为我们两国之间深厚友谊的象征。”随后，威廉·高伯拉瓦和总理班夫人与帮助建设这座大厦的中国工程技术人员亲切握手，并在中国大使的陪同下参观了大厦的模型和图片展览。

在开工仪式上，总理班夫人为“班厦”破土，表示大厦工程正式开工兴建。

25日晚，马子卿大使举行招待会，庆祝大厦开工。总理班夫人出席了招待会。

On November 24, 1970, the Sri Lankan government held a grand commencement ceremony for Bandaranaike Memorial International Conference Hall (BMICH), a symbol of Sino-Sri Lanka friendship. William Gopallawa, Governor-General of Sri Lanka and Prime Minister Sirimavo Bandaranaike attended the ceremony. The governor-general delivered a speech after reviewing the Sri Lankan army honor guard. He expressed gratitude towards the Chinese government in aiding Sri Lanka in building BMICH, and stated, "BMICH will become a symbol of friendship between the two countries." Prime Minister Sirimavo Bandaranaike asked the Chinese ambassador to convey her thanks to the Chinese people, especially to Premier Zhou Enlai. Chinese Ambassador Ma Ziqing highly valued the contributions the late Prime Minister Bandaranaike had made in fighting against imperialism and colonialism, defending national independence and promoting friendly relations between the two countries. Ambassador Ma said, "Through our joint efforts, BMICH will become a symbol of deep friendship between China and Sri Lanka." Later, William Gopallawa and Prime Minister Sirimavo Bandaranaike shook hands with Chinese engineers involved in BMICH construction, and inspected the BMICH model and presentation.

Prime Minister Sirimavo Bandaranaike broke ground for BMICH project at the commencement ceremony, indicating the official start of the project.

On the evening of November 25, 1970, Ambassador Ma Ziqing held a reception to celebrate the project's commencement. Prime Minister Sirimavo Bandaranaike attended the event.

总理班夫人在大厦工地破土（在大厦开工典礼时）

Prime Minister Sirimavo Bandaranaike broke ground at the construction site of BMICH commencement ceremony

特使徐向前
Special Envoy Xu Xiangqian

1973年5月16日上午，中华人民共和国特使徐向前（以后以“特使”做亲切称呼）乘专机到达科伦坡，参加“纪念班达拉奈克国际会议大厦”的揭幕典礼。在机场上受到斯里兰卡总理班夫人的热烈欢迎。当特使下飞机时，总理班夫人上前同他热烈握手，这时，乐队演奏了欢迎乐曲。

到机场欢迎的还有：斯里兰卡政府官员、中国驻斯里兰卡大使黄明达和大使馆官员、修建“班厦”的中国技术组成员。

在这次参加“班厦”揭幕典礼期间，特使还参谒了前总理班达拉奈克的墓。班达拉奈克总理的墓身是一块巨大的不规则的黑色大理石，周围五根巍然屹立的柱子象征和平共处五项原则，见了很让人震动，觉得有一种浩然正气在空中回荡。

1973年5月，特使在“班厦”的几位主要负责人陪同下去了康提。当地驻军司令招待特使时，用的是一盘中国国内罕见的，斯里兰卡也很珍贵的人参果做的招待。并打开了佛牙寺珍藏佛牙的金塔，请特使瞻仰。

On the morning of May 16, 1973, special envoy of the People's Republic of China Xu Xiangqian (short as the special envoy) arrived in Colombo by a special plane to attend the Opening Ceremony of BMICH. Prime Minister Sirimavo Bandaranaike warmly welcomed the special envoy and shook hands with him when he disembarked from the plane, and a band played welcoming music.

Sri Lankan government officials, Chinese ambassador to Sri Lanka Huang Mingda, embassy officials and Chinese technical team members of the BMICH project also welcomed the special envoy.

The special envoy paid homage at the cemetery of the late Sri Lankan Prime Minister Bandaranaike. The tombstone is made of a huge block of irregular black marble, and the surrounding five majestically standing columns symbolize the Five Principles of Peaceful Coexistence. Visitors were surprised by the righteousness in the atmosphere.

Accompanied by several principals of BMICH, the special envoy visited Kandy in May 1973. The local garrison commander treated the special envoy with ginseng fruit, which is rare in both China and Sri Lanka, and invited the special envoy to see a gold pagoda in the Temple of the Tooth.

1. 总理班夫人热烈欢迎徐向前特使
Prime Minister Sirimavo Bandaranaike warmly welcomed special envoy Xu Xiangqian

2. 徐向前特使在机场接受花环
Special envoy Xu Xiangqian garlanded at the airport

3. 徐向前特使在机场会见内阁官员
Special envoy Xu Xiangqian met Cabinet officials at the airport

4. 徐向前特使在机场在金书上签名
Special envoy Xu Xiangqian signing the Golden Book at the airport

1

2

3

4

1. 徐向前特使与斯里兰卡总理班夫人在拉塔马拉纳机场
Special envoy Xu Xiangqian and Prime Minister Sirimavo Bandaranaike at Ratmalana Airport

2. 总理班夫人与徐向前特使在招待会上
Prime Minister Sirimavo Bandaranaike and Special envoy Xu Xiangqian at the reception

3. 在豪拉构腊班达拉奈克墓地
At Bandaranaike Cemetery in Horagolla

1

2

3

竣工典礼
Completion Ceremony

竣工揭幕典礼由总理班夫人亲自主持，总督威廉·高伯拉瓦在仪式上讲了话，他盛赞："这座大厦的建成将成为我们两国之间深厚友谊的象征！"并与徐向前特使一道在礼仪大厅点亮传统油灯。

典礼仪式结束后，总理班夫人将"班厦"竣工典礼仪式和在会议大厅上演的，由总理班夫人女儿（后曾任斯里兰卡总统）编导的名为《光明之路》的舞剧（这个舞剧描写的是历史上从中国僧人法显赴斯里兰卡，一直到兴建和完成"纪念班达拉奈克国际会议大厦"期间中斯两国人民之间的传统友谊）拍成的电影纪录片及一个底座纵34.5厘米，横34.5厘米，高17.5厘米、镶有当地产的红宝石的银质"班厦"模型赠送给周恩来总理。"班厦"模型现标明为"对外援助的典范——'班厦'模型"收藏在中国国家博物馆亚洲礼品馆。

Prime Minister Sirimavo Bandaranaike presided over the Completion and Unveiling Ceremony and Governor-General William Gopallawa gave much praise, saying, "BMICH will be a symbol of China-Sri Lanka friendship!" William Gopallawa and the special envoy Xu Xiangqian jointly lit the traditional oil lamp at Etiquette Hall.

After the ceremony, a drama called Path to Brightness written and directed by the daughter of Prime Minister Sirimavo Bandaranaike (who later served as Sri Lankan president) was shown at the Completion Ceremony and Conference Hall, the drama records the traditional friendship between the Chinese and Sri Lankan people since the Chinese monk Fa Xian went to Sri Lanka until the construction and completion of BMICH. The drama was made into a film documentary and Mrs. Bandaranaike sent it and a silver BMICH model set with local rubies (the base size is 34.5cm wide, 34.5cm long and 17.5cm high) to Premier Zhou Enlai. The BMICH model has become an "Example of China-aided Projects——the BMICH model is now at the Asia Gift Pavilion of National Museum of China.

总理班夫人在竣工典礼上致欢迎词
Prime Minister Sirimavo Bandaranaike delivered a welcome address at the completion ceremony

1. 中国特使回复致词
The Chinese special envoy made a speech at the ceremony

2. 竣工典礼仪式上总理班夫人在主持传统的煮牛奶仪式
Prime Minister Sirimavo Bandaranaike hosted the traditional milk-boiling rite at the completion ceremony

1. 竣工典礼仪式上中国特使徐向前在打开椰花鞘
The Chinese special envoy opened the coconut flower sheath at the completion ceremony

2. 中国特使徐向前与斯里兰卡总督威廉・高伯拉瓦在礼仪大厅处点亮传统油灯
The Chinese special envoy and Governor-General William Gopallawa lit the traditional oil lamp at Etiquette Hall

3. 总理班夫人送给周恩来总理的银质“班厦”模型
The silver BMICH model Prime Minister Sirimavo Bandaranaike sent to Premier Zhou Enlai

极大鼓舞
Great Inspirations

在“班厦”施工期间，两国领导人都非常关心工程进展的情况，总理班夫人和其他斯里兰卡政府官员，还多次亲自到工地参加劳动，给予“班厦”所有的工作人员以极大鼓舞。

During the construction period, the leaders of the two nations were highly concerned about the project's progress. Prime Minister Sirimavo Bandaranaike and other Sri Lankan government officials visited the BMICH construction site many times, which greatly inspired the workers.

总理班夫人和官员们在“班厦”建设的现场
Prime Minister Sirimavo Bandaranaike and other Sri Lankan government officials inspected BMICH construction site

1. 总理班夫人和部长们在“班厦”工地现场
Prime Minister Sirimavo Bandaranaike and Sri Lankan ministers at the BMICH construction site

2. 1972年6月7日总理班夫人到工地视察时，同工地的中国工程技术人员和工人一起合影。前排左一为使馆二秘陈玉妮，左二为总理班夫人，左三为大使马子卿，左四为工程组长郝辅堂，二排左一为工程组副组长兼总工程师由宝贤
Prime Minister Sirimavo Bandaranaike took a group photo with Chinese engineers at the BMICH construction site on July 7, 1972.Front Row: 1st from Left: Second Secretary of Embassy Chen Yuni, 2nd from Left: Prime Minister Sirimavo Bandaranaike, 3rd from Left: project team leader Hao Futang.Second Row: 1st from Left: Deputy project team leader and chief engineer You Baoxian

3. 总理班夫人在视察施工现场
Prime Minister Sirimavo Bandaranaike visited the construction site

亲切关怀
Loving Care

一 宴请

在大厦完工前夕，一些工作人员陆续回国，留在工地的施工人员大约有200多人。这天，总理班夫人在总理府的院子里，宴请所有的施工人员及使馆人员。宴会开始时，总理班夫人用英语讲话，言谈既亲切又有力。她说道："我们国家政府最高领导人请客吃饭，规定最多不能超过150人，而我今天请了200多人，我是破了国家的一些规定，但我相信，如果我向全国人民讲，我请的是'班厦'工程的技术人员，他们为我们大厦工程做了贡献，为我们斯中友谊也做了贡献，全国人民都会支持、原谅我的，不会说我做得不对。"

吃完饭临走时，总理班夫人亲自向每个人赠送了一个纪念瓷盘，瓷盘上方写着"纪念班达拉奈克国际会议大厦"，中间是精致宏伟的大厦形象，下面则写着"斯中友谊的象征"。文字全部用僧伽罗语和汉语双语书写，直径约20cm，制作非常精美。

二 关怀

为了庆祝开工两周年，工地决定会餐一次，那天正是周六。忙了一天的工作，人们都感觉累。晚上会餐时，大家还是很高兴，有吃有喝有笑，可到了晚上十点，许多人开始呕吐，还有不少人腹泻。总共有200多人都被立即送往医院。

第二天一早，总理班夫人知道此事后，立即派卫生部部长和工作人员到工地来询问情况，问大家都吃了些什么，还到厨房把剩余的食品取样带走化验。经化验是吃的茶叶蛋坏了。总理班夫人怕是有人下毒，在医院都派了兵。大家也很紧张。

(I) Banquet

Prior to the project's completion, Chinese staff returned home successively, leaving behind about 200 construction personnel at the construction site. One day, Prime Minister Sirimavo Bandaranaike invited all construction staff and those at the embassy to have dinner in the yard of the Prime Minister's Office. Before dining, Prime Minister Sirimavo Bandaranaike spoke in English kindly and powerfully. She said, "According to Sri Lankan regulations, top national leaders can treat at most 150 people, but today I am hosting over 200 people. Although I am breaching the national rule, but I believe the Sri Lankan people will support and forgive me if I told them that I treated technical staff for the 'BMICH' project who have made a great contribution to the project's construction and Sri Lanka-China friendship."

After the dinner, Prime Minister Sirimavo Bandaranaike presented everyone an exquisite memorial porcelain dish that is 20cm in diameter, with "Bandaranaike Memorial International Conference Hall" on the top, with an elegant and magnificent illustration of the hall in the middle and "Symbol of Sri Lanka-China Friendship" at the bottom. All the script on the dish are written in both Sinhala and Chinese.

(II) Caring

One Saturday, workers had a dinner party at the construction site to celebrate the second anniversary of the project's commencement. The workers were tired after a hard day's work, and they were delighted to have dinner together. But at 10p.m., many workers began to vomit while others had diarrhea, so all 200 workers were sent to hospital immediately.

The next morning, Prime Minister Sirimavo Bandaranaike learned about this, and she promptly sent the health minister and staff to the construction site, where they asked questions and took samples from the leftover food in kitchen for testing. The test results indicated the eggs were spoiled. Prime Minister Sirimavo Bandaranaike sent soldiers to hospital for fear of poisoning. Everyone was nervous.

总理班夫人赠送的纪念瓷盘

Commemorative porcelain dish sent by Prime Minister Sirimavo Bandaranaike

第三天，周恩来总理从《参考消息》报上得知此事，马上批评大使为什么没马上报国内，并问是否需派医生过去，或派专机把病重的人接回国内治疗。周总理对在“班厦”工作的人们关心得是无微不至，使得全体人员感动万分。

当时医院床位紧张，但一听说是建大厦的中国工人生病，当地病人都主动让出床位，让中国人住院治疗。

住院期间，人们得到了两国总理的亲切关怀和斯里兰卡人民的热情帮助，使工程技术人员和工人都非常感动。

On the third day, Premier Zhou Enlai learned about this from Reference News, and he immediately criticized the ambassador for not reporting this incident to the Chinese government, and asked whether it was necessary to send Chinese doctors to Sri Lanka or to bring the seriously ill workers back to China by special plane. All the staff working on the BMICH project was deeply touched by Premier Zhou's meticulous concern for their well being.

Many local patients volunteered their beds to Chinese workers as hospital beds were in short supply.

While in hospital, the Chinese workers were moved by the loving care of the Sri Lankan Prime Minister and Chinese Premier as well as the enthusiastic assistance of the Sri Lankan people.

友好“使者”——“米杜拉”
Friendly "Envoy"—— "Mithura"

1972年6月24日，应中华人民共和国政府的邀请，斯里兰卡共和国总理西丽玛沃·班达拉奈克夫人来我国进行国事访问。国务院总理周恩来等，前往机场热烈欢迎贵宾。毛泽东主席在中南海接见了总理班夫人。

这次访问，总理班夫人还带来一份特殊的礼物，一头名叫“米杜拉”的小象。

1972年6月28日在首都体育馆大厅里，举行了赠象仪式。周恩来总理亲自出席，并在仪式上讲话。总理班夫人在仪式上致词，她说：“‘米杜拉’将成为我们两国儿童之间友谊的一个活的象征。”

On June 24, 1972, at the invitation of the government of the People's Republic of China, Sri Lankan Prime Minister Sirimavo Bandaranaike paid a state visit to China. Premier Zhou Enlai and other Chinese officials warmly welcomed Prime Minister Sirimavo Bandaranaike at the airport and Chairman Mao Zedong met Prime Minister Sirimavo Bandaranaike in Zhongnanhai.

During her visit in China, Prime Minister Sirimavo Bandaranaike brought a special gift – an elephant named "Mithura".

On June 28, 1972, an elephant presentation ceremony was held in the hall of Capital Indoor Stadium. Chinese Premier Zhou Enlai attended the ceremony and delivered a speech. Prime Minister Sirimavo Bandaranaike addressed the audience, saying, "'Mithura' will be a lively symbol of friendship between children of the two countries."

周恩来总理和西丽玛沃·班达拉奈克总理在机场亲切握手
Sri Lankan Prime Minister Sirimavo Bandaranaike and Chinese Premier Zhou Enlai shaking hands at the airport

1. 1972年6月28日晚，毛泽东主席在中南海会见斯里兰卡总理西丽玛沃·班达拉奈克夫人时情形
Chairman Mao Zedong met Prime Minister Sirimavo Bandaranaike in Zhongnanhai on the evening of June 28, 1972

2. 西丽玛沃·班达拉奈克总理将小象“米杜拉”交给周恩来总理和六名中国儿童代表
Sri Lankan Prime Minister Sirimavo Bandaranaike presented "Mithura" to Chinese Premier Zhou Enlai and six Chinese children

3. 可爱的“米杜拉”在人们热烈的簇拥下走进北京动物园
The lovely "Mithura" enters the Beijing Zoo surrounded by an enthusiastic crowd

4. 中国白唇鹿
Chinese white-lipped deer

总理班夫人代表斯里兰卡儿童送给中国儿童的小象名为“米杜拉”，“米杜拉”在僧伽罗文中是“朋友”的意思。小象穿着色彩鲜艳、富有斯里兰卡传统艺术特色的衣裳。周恩来总理宣布为感谢斯里兰卡儿童的友情，中国儿童将赠送一对白唇鹿给斯里兰卡儿童。于9月7日，在斯里兰卡首都科伦坡德希瓦纳动物园举行了中国政府赠送斯里兰卡儿童白唇鹿交接仪式，中国驻斯里兰卡大使参加了仪式。

可爱的“米杜拉”随后被送到北京动物园饲养。

The small elephant sent by Prime Minister Sirimavo Bandaranaike on behalf of Sri Lankan children is called "Mithura", which means "friend" in Sinhala. "Mithura" wears a colorful cloth with rich traditional art designs. Premier Zhou Enlai declared that Chinese children would present a pair of white-lipped deer to Sri Lankan children as a token of thanks. The white-lipped deer handover ceremony was held in Dehiwala Zoo in Colombo, capital of Sri Lanka on September 7, 1972. The Chinese ambassador to Sri Lanka participated in the ceremony.

The lovely "Mithura" was then sent to the Beijing Zoo.

密切合作
Close Cooperation

运输钢屋架的车进入“班厦”工地现场时的情景
Vehicle carrying a steel roof truss enters the BMICH construction site

为了保证钢屋架檩托准确就位，钢屋架不是现场拼装，而是将整个屋架分3段或4段发运。这样发运的钢屋架每件的长度为10余米，有的最高点达4.8米。当时，从码头到工地的路上虽没有什么桥涵，却有电线的高度限制，使得运输钢屋架的车无法通过。斯里兰卡政府就让电力局在前面开路，遇到电线电缆，电力局先把电线剪开，运钢屋架的车过去以后，电力局再马上把电线接上。就这样，双方密切配合，把钢屋架全部运到了工地上。

To ensure the steel roof truss purlin hanger was placed in the correct position, the steel roof truss was not assembled on site; instead, it was transported in three to four sections, so the steel roof truss was about 10 meters long, sometimes as high as 4.8 meters. Although there were no culverts from the dock to the construction site, the height of the electrical wires along the route limited the vehicle's height. Therefore, the Sri Lankan electricity bureau cut off the electrical wires to allow vehicles carrying the steel roof trusses to pass, and then reconnected them later. Through close cooperation, all the steel roof trusses were transported to the construction site.

工作友谊
Friendship Established at Work

在建设大厦期间，两国工作人员结下了深厚的友谊，他们一同劳动，携手建设这个大厦。

在施工过程中，斯里兰卡方面的总工程师达莫德林嘎姆（Damodlingam）和我国专家接触最多，与我国专家合作得非常融洽，在技术上也常互相切磋，互相探讨。他和他们的副局长有时会请中国的主要技术人员到家中做客，中国专家有时也会请他和其他斯方工作人员到工地来就餐，一起谈论工作，交流生活经历。整个工程完工后，国内又运来大量餐具供大厦使用，有刀、叉、盘子、碗，一共5万多件。他与我国负责人一起在大厦地下室清点了三天，他负责点数，中方人员负责做记录。建设“班厦”的日日夜夜，让两国工程技术人员结下了难以忘怀的友谊。

Worker from the two countries joined hands in constructing BMICH, and established deep friendship.

During construction, BMICH chief engineer Mr. Damodlingam contacted and cooperated well and frequently with Chinese experts. They often discussed technical problems. Mr. Damodlingam and the deputy director general sometimes invited main Chinese technical personnel to their homes, and the Chinese experts would invite Mr. Damodlingam and other Sri Lankan staff to dine at the construction site, during which they discussed work and life experiences. After the project was completed, the Chinese government sent 50,000 pieces of dinnerware including knives, forks, plates and bowls. Mr. Damodlingam and the Chinese principals checked the dinnerware together for three days in the basement of BMICH, with Mr. Damodlingam counting and the Chinese personnel recording the numbers. The technical personnel of the two countries established an unforgettable friendship during the BMICH construction period.

前排左起第五为斯方大厦总工程师达莫德林嘎姆，第六为工程组副组长兼总工程师由宝贤，第八为工程组的施工组长王茂堂

Front Row: Fifth from left: Damodlingam, chief engineer of BMICH. Front Row: Sixth from left: You Baoxian, deputy project team leader and chief engineer. Front Row: Sixth from right: Wang Maotang, construction team leader

后来，我国还请他和工程部常务秘书、正局长、副局长、总工程师一起来到中国，由中方工程负责人陪他们参观访问了上海、南京、苏州、杭州、长沙和广州等地。

当地工人们都讲僧伽罗语，但我们派到现场的是英语翻译，根本不会僧伽罗语，在工作和生活中很难交流。后来我们就找了两个当地华侨，帮助做僧伽罗语翻译。他们翻译生活上的事情没问题，但由于不懂建筑技术，不能掌握、理解技术上的词汇，也不知道该怎样翻。

工人们只好连说带比划。后来，斯里兰卡工人教中国工人说僧伽罗语，中国工人教他们说汉语，还教他们建筑技术。就这样，两国工人之间相互交流技术、交流语言、一起工作，生活上相互关心，结下了真挚深厚的友谊。

在一起工作期间，当地工人经常在家中做一些他们的拿手菜，带到工地给中国工人吃。中国工人病了，他们都到宿舍看望。他们受了伤，中国工人也非常关心，帮忙把他们送到医务室治疗。后来中国工人回国时，很多人都流下了难离难舍的眼泪。

我国派出的都是技术顶尖的工人。就拿木工陶金标师傅来讲，他教会了斯里兰卡木工使用中国的木工机械和手工木工工具，提高了他们的木工技能，培养了人才，也使得大厦施工的进度加快了。

The Chinese government invited Mr. Damodlingam, the permanent secretary, director, deputy director to visit Shanghai, Nanjing, Suzhou, Hangzhou, Changsha and Guangzhou, with the chief engineer of the engineering department of China, and Chinese engineering principals accompanying them.

Local workers spoke Sinhala, but the on-site English interpreter didn't know Sinhala, making communication difficult in work and life, so two local overseas Chinese were invited to interpret, but they were only good at translating things about life, as they neither knew architectural technology nor understood or mastered technical vocabulary, let alone interpretation.

Sri Lankan workers spoke with gestures, and they taught Chinese workers Sinhala, while Chinese workers taught them Chinese and architectural technology. Workers from the two countries exchanged technologies and languages, worked together and cared about each other in life, thus establishing deep and sincere friendship.

Local workers often brought with their specialty dishes to the construction site for the Chinese workers to taste, and visited them when they were sick in the dormitory; the Chinese workers also cared a lot about their Sri Lankan counterparts, and sent injured Sri Lankan workers to medical clinics for treatment. When the Chinese workers returned to China, many Sri Lankan workers shed tears.

The Chinese government sent workers with top technical expertise for the project. For instance, Tao Jinbiao, a carpenter, taught his Sri Lankan counterparts how to use Chinese woodworking machinery and tools, thus elevating their carpenter skills, cultivating talents and speeding up construction.

有个木工，是十来个斯里兰卡木工的领头人。他很聪明，学技术又快又好。大家相处久了，就都用他的昵称称呼他“古伦巴拉”。斯里兰卡总工程师达莫德林嘎姆都知道他的成绩，还给他增加了工资。陶金标师傅回国时，古伦巴拉带着全家人，送了些水果给陶金标师傅，还送给陶师傅一张他的照片留念，陶金标师傅至今仍珍贵地保存着。

The head of the Sri Lankan carpenters is named T. B. Wickramasinghe, he was smart and mastered technology quickly and well. After they got on well with each other, they called him by his nickname "Kurunpara". After knowing this, Chief Engineer Mr. Damodlingam raised his salary. When Tao Jinbiao returned to China, Wickramasinghe and his family members sent fruits and a photo of him to Tao Jinbiao. He still has the precious photo today.

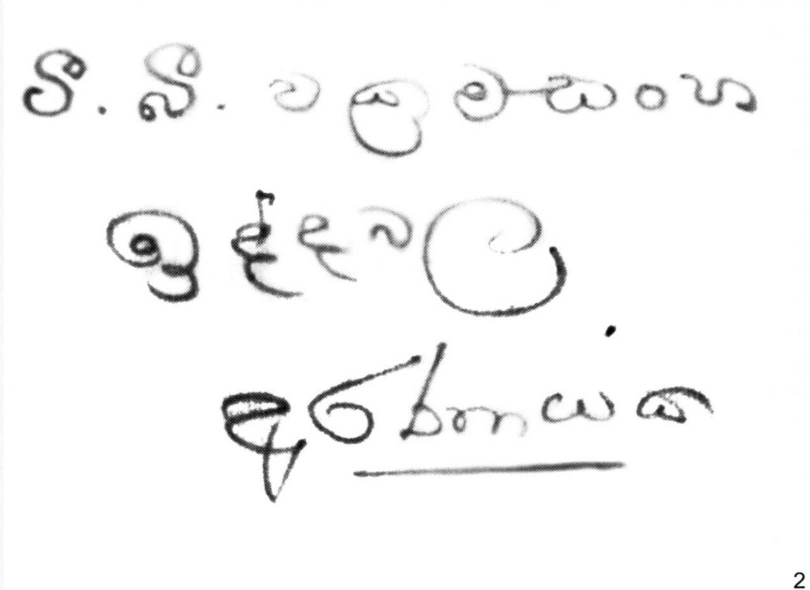

1. 古伦巴拉先生的照片
A photo of Mr. T. B. Wickramasinghe

2. 古伦巴拉先生在照片背后的留言，留言内容是他的家庭住址及他的签名。
The message written by Mr. T. B. Wickramasinghe on back of the photo, indicating his home address and his signatue

在会客室给斯方工程师一家放电影
A film shown for Sri Lankan engineer's family in the reception room

一 观看彩排

施工后期的一天，总理班夫人邀请项目主要负责人员、使馆参赞和其他领导，还有几个设计人员，一起观看她女儿编导的为竣工典礼准备的舞剧《光明之路》的彩排。

看完彩排后，总理班夫人说："我请你们吃饭。"吃的饭是每人一包用不知是什么树的叶子包得像礼包一样很精致的米饭，打开一看，里面是盖浇饭，饭上面的菜有咖喱牛肉、土豆、萝卜。

(I) Dress rehearsal

One day during the later stages of construction, Mrs. Bandaranaike invited the main principals, the embassy counselor and other leaders, as well as several designers to watch the dress rehearsal of the drama Path to Brightness directed by her daughter for the completion ceremony.

After watching the rehearsal, Mrs. Bandaranaike said: "I am inviting you to dinner." She gave each person one pack of rice wrapped with a leaf, and inside was rice served with vegetables much like curry beef, with potatoes and turnips on the top.

二　斯方客人

施工期间有许多斯里兰卡的客人来参观，有的还说想看中国电影，工地负责人就安排给他们放电影。当时，是在“文化大革命”期间，许多电影遭封杀，只有四部从国内带过去的电影来回放，被称为“三战一打击”，“三战”是《地道战》、《地雷战》、《南征北战》，“一打击”是《打击侵略者》。电影是汉语原声片，斯里兰卡的客人们也听不懂，但因为都是战争片，他们也看得津津有味，笑声不断。

三　国内客人

施工期间，大使馆还安排了许多中国代表团来大厦工地参观，一起交流学习，让国内的客人们感受一下，中国援外工作人员在这么炎热的天气里，是怎样为援外事业而努力工作的。国内的客人也给远在斯里兰卡的援外工作者带来了祖国的形势和新闻，还有很多难忘的欢笑。

当时有国家体委办公厅主任带队的乒乓球代表团，团内有李富荣、梁丽珍等男女运动员，羽毛球代表团有侯加昌、陈玉娘等人。还有武汉杂技团，带队的是副团长夏菊花。这些代表团主要是参观大厦的施工。有的还会给施工人员做些表演，例如乒乓球代表团就在食堂里表演了他们的精彩球技；不能表演的就在一起座谈，工地的同事们对他们的招待也非常热情周到。

(II) Sri Lankan guests

Many Sri Lankans visited the BMICH construction site wanted to see Chinese films, so the head of the construction site presented movies for them. Due to historical reasons, many of the films were not allowed to be shown, and only four from China were available. They were: Three Wars and One Fight, Tunnel Warfare, Landmine Warfare and Fight North and South, Fight the Invaders. Although the Sri Lankan side could not understand the original Chinese films, they watched films with relish and laughed from time to time.

(III) Chinese delegations

The embassy arranged for Chinese delegations to visit the BMICH construction site for exchanges, and let the Chinese guests empathize with how China-aided project workers work hard in such hot weather. The Chinese delegations also brought the situation and news from China and many memorable laughs.

Chinese delegations included a table tennis one led by the director of the General Office under the State Physical Culture and Sports Commission, with members including Li Furong and Liang Lizhen, a badminton delegation with members of Hou Jiachang and Chen Yuniang, and the Wuhan Acrobatic Troupe led by deputy head Xia Juhua. These groups mainly visited the building construction site, and sometimes they performed for the workers. For example, the table tennis delegation demonstrated their excellent skills; sometimes, the delegations met with workers and were warmly treated.

“班厦”项目施工主要负责人和参观“班厦”工地的国内代表团的合影

A group photo of the principals on BMICH site and the visiting Chinese delegation

报纸剪辑

Newspaper Clippings

一　人民日报

在“班厦”即将落成时，中斯双方政府报刊都发表了重要报道。我国在1972年7月13日人民日报第六版以“斯里兰卡国民议会，通过西丽玛沃·班达拉奈克总理政策声明，重申奉行不结盟政策，并努力使印度洋成为和平区”的醒目标题发表报道。

(I) People's Daily

The Chinese and Sri Lankan government newspapers released important reports on the completion of BMICH. On July 13, 1972, the People's Daily released on the sixth page a report with the eye-catching headline of "The Sri Lanka National Assembly passes policy statement made by Prime Minister Sirimavo Bandaranaike, reaffirming its adherence to the nonalignment policy and striving to make the Indian Ocean a Peaceful Region".

1. 1973年5月17日人民日报报道特使徐向前到达科伦坡参加“班厦”揭幕典礼
   The People's Daily reported on May 17, 1973 that Chinese special envoy Xu Xiangqian arrived in Colombo to attend the BMICH unveiling ceremony
2. 1972年7月13日人民日报
   The People's Daily on July 13, 1972

1973年5月17日　星期四　第五版

我国特使徐向前到达科伦坡

斯里兰卡总理班达拉奈克夫人到机场热烈欢迎

参加「纪念班达拉奈克国际会议大厦」揭幕典礼

新华社科伦坡一九七三年五月十六日电　中华人民共和国特使、人大常委会副委员长徐向前五月十六日上午乘专机到达科伦坡，参加“纪念班达拉奈克国际会议大厦”的揭幕典礼。

徐向前特使在机场上受到斯里兰卡总理班达拉奈克夫人的热烈欢迎。

在洋溢着热烈友好气氛的机场上，斯中两国国旗迎风飘扬。

当徐向前特使走下飞机时，总理班达拉奈克夫人上前同他热烈握手。这时，乐队演奏了欢迎乐曲。

从机场到徐向前特使下榻的总理官邸的路上，许多人挥动斯中两国国旗和彩旗，热烈欢迎中国客人。

今天到机场欢迎的有：斯里兰卡政府的许多部长和副部长、陆军司令和其他高级军政官员以及斯中友好协会主席瓦尔皮塔等。

中国驻斯里兰卡大使黄明达和大使馆官员、修建“纪念班达拉奈克国际会议大厦”的中国技术小组成员以及华侨代表，也到机场迎接。

徐向前特使的随行人员以及陪同他们来斯里兰卡的斯里兰卡驻中国大使卡朗纳戈达和夫人同机到达。

图为“纪念班达拉奈克国际会议大厦”　新华社记者摄

「纪念班达拉奈克国际会议大厦」

中国斯里兰卡深厚友谊的象征

斯里兰卡政府和人民为纪念已故总理所罗门·班达拉奈克而修建的“纪念班达拉奈克国际会议大厦”，两年多来通过斯中两国工人和技术人员的密切合作和共同努力，已于一九七三年四月全部竣工，五月十七日举行揭幕典礼。

一九五六年，以班达拉奈克为首的人民联合阵线在大选中获胜，成立政府，班达拉奈克任总理。新政府成立后，在内外政策方面采取了一系列维护民族利益的措施。它对外奉行独立自主、和平中立的不结盟政策，拒绝参加军事集团，并撤销了外国在斯里兰卡的军事基地；在国内着手建立民族工业，发展民族经济和文化，规定僧伽罗语作为官方用语，等等。

一九五七年一月，周恩来总理应邀访问斯里兰卡，同班达拉奈克总理举行会谈。两国总理确认以和平共处五项原则和万隆会议十项原则作为指导两国关系的准则，并在这一准则指导下，正式建立了外交关系。从此，中斯两国政府和人民间的友好合作关系不断得到巩固和发展。

班达拉奈克总理坚持反对帝国主义和殖民主义、捍卫民族独立的斗争，并对发展斯中两国间的友好关系作出了积极贡献。为了纪念已故的班达拉奈克总理，一九六四年中斯两国政府商定由中国帮助斯里兰卡建筑一幢国际会议大厦。一九六五年三月，由总理班达拉奈克夫人亲自主持了奠基典礼。一九七〇年十一月二十四日，“纪念班达拉奈克国际会议大厦”正式动工。我国派出工人和技术人员，同斯里兰卡工人和技术人员共同劳动，一起工作。这项工程得到总理班达拉奈克夫人和斯里兰卡政府的大力支持和关怀。为了领导大厦的建设工作，斯里兰卡内阁还成立了专门机构。班达拉奈克夫人曾多次视察工地，并和其他内阁官员一起参加义务劳动，给斯中两国工人和技术人员以很大鼓舞。

“纪念班达拉奈克国际会议大厦”包括一个有一千五百个座位的大会堂、九十个代表团的办公室、六个委员会会议室和秘书处等部分，总建筑面积达三万二千多平方米。在大会堂门厅内，有一座用白色大理石雕刻的已故总理班达拉奈克半身像。门厅的尽端，有一幅表现斯里兰卡美丽富饶的国土和勤劳勇敢的人民的大油画。油画后面是宴会厅。楼上就是大会堂。这座大厦将成为斯里兰卡举行重要政治活动和召开国际会议的场所。

在建筑大厦的过程中，中斯两国工人和技术人员一起工作，相互学习，彼此间结下了深厚的友谊。正象斯里兰卡共和国总统威廉·高伯拉瓦在开工典礼上所说的那样：“这座大厦的建成将成为我们两国之间深厚友谊的象征。”

国际资料

1

斯里兰卡国民议会

通过西丽玛沃·班达拉奈克总理政策声明

重申奉行不结盟政策，并努力使印度洋成为和平区

据新华社科伦坡一九七二年七月十一日电　斯里兰卡共和国国民议会最近通过了西丽玛沃·班达拉奈克总理六月二十三日在国民议会上发表的政策声明。

声明说：“已故的所·韦·里·迪·班达拉奈克先生所制定的、我的政府一贯奉行的不结盟政策得到了所有各国的尊重和谅解。”

声明说，在一九七三年，当目前正在中华人民共和国援助下建造的纪念班达拉奈克大厦落成后，“斯里兰卡能成为招待一次不结盟国家最高级会议的东道国。”

班达拉奈克夫人的声明说：“在去年的联合国大会上，我自从一九六四年开罗会议以来就提出的宣布印度洋为和平区的基本思想被采纳了，我将在这方面坚持不懈地作出努力。”

声明说：“一九七一年十一月，我向前众议院提出了一项详细的五年计划，该计划提出了我的政府在未来五年中的经济方针。”这项计划包括了增加出口，提高大米生产，禁止进口干辣椒、洋葱和其他副食品，改进家畜饲养工作，提高椰子种植的标准，通过谈判达成一项长期的国际茶叶协定来确保比较稳定的茶叶价格等。

2

二　斯方剪报

(II) Sri Lankan newspaper clipping

斯里兰卡相关报刊剪辑一

Sri Lankan newspaper clippings I

欢迎宋庆龄副主席周恩来总理

හ්වාන් යිං සුං චිං ලිං ෆු චු ෂී චූ එන්ලායි ත්සුං ලි

උප ජනපතිනි සූං හා අගමැති චූ සංගනිමු

චීන ජනරජයේ සුං චිං ලිං උප ජනාධිපතිනිය සහ අගමැති චූ එන්ලායි, උප අගමැති මහ සෙනෙවි චෙන් යි ඇතුළු තිදෙනා හතලිස් හය දෙනකුගෙන් යුත් සහපිරිවරින් අද සවස 2.20 ට විශේෂ ගුවන් යානයකින් රත්මලානේ ගුවන් තොටුපොළින් ලංකාවට සම්ප්‍රාප්ත වෙති.

ඉංදු-චීන අවුල ගැනත් කතාබස්

හඳුන්වා දීම

චීනඅමුත්තෝ: විශේෂ ලිපි

උප ජනපතිනි සු. චිං. ලිං　　අගමැති චූ එන් ලායි　　උප අගමැති චෙන් යි

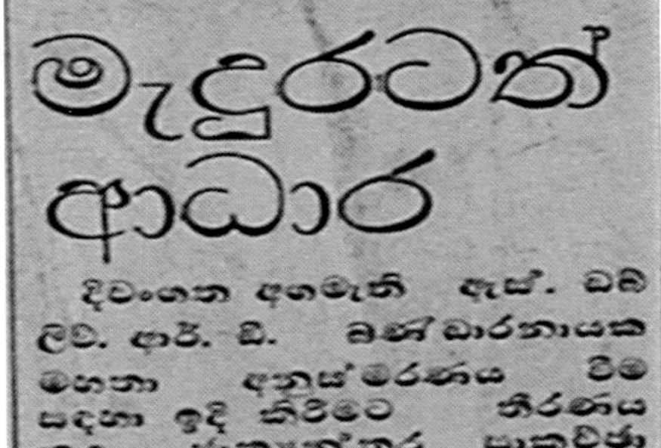

සාකච්ඡා මැදුරටත් ආධාර

දිවංගත අගමැති එස්. ඩබ්ලිව්. ආර්. ඩී. බණ්ඩාරනායක මහතා අනුස්මරණය වීම සඳහා ඉදි කිරීමට තීරණය කළ ජාත්‍යන්තර සාකච්ඡා මන්දිරයට ආධාරදීමට චීන මහජන රජයේ අගමැති චූ එන් ලායි මහතා පොරොන්දු වී ඇත.

Chinese help build hall

(By a "Daily News" reporter).

The Bandaranaike Memorial International Conference Hall—work on which has already begun—will be able to accomodate 540 delegates from 90 nations and will also provide offices and all amenities and facilities for international conferences.

According to a Press communique issued yesterday by the Ministry of Public Works the preliminary work on the construction of this Hall which was promised by Mr. Chou-en-Lai, Prime Minster of China as a gift, has been begun by a team of architects and engineers from China with the assistance of the PWD

This building which is estimated to cost Rs. 19 milion is coming up at the site of the Havelock Golf Course, Bullers Road, Colombo.

SWRD Hall: OK for new plan

CHINA TO USE HER MARBLE FOR RS. 25M. BUILDING

THE Ceylon Government's amended plan for the Bandaranaike Memorial Hall has been accepted by the People's Republic of China.

The Government has proposed the construction of a hall-cum-theatre and hotel at an estimated cost of about Rs. 25,000,000.

The original plan was for an elaborate conference hall that would accommodate 540 delegates and 110 observers with a main auditorium and balconies.

The hall now planned will be a simpler structure which will however, accommodate about 800 persons.

The new suggestions of the Government were conveyed by members of a delegation led by Mr. V. C. de Silva, Director of Public Works, which visited Peking in this connexion last month.

It is learnt that marble from China will be used for the construction of the building.

It has been suggested that the hotel should include a swimming pool.

Chinese monument

WE have said it before, and we think it necessary to say again, with emphasis, that the Sirima Bandaranaike Government is not doing the memory of Mr. S. W. R. D. Bandaranaike any honour by having a foreign government put up here, in our own country, a monument to him. We urge that a number of very good reasons now demand that the Government cry halt to the project and reconsider the whole matter.

It was a shameless statement Dr. N. M. Perera made on the subject in his budget speech last July: "An International Conference Hall has been one of the more urgent needs of Ceylon in view of the increasing demand for Ceylon as a venue of international conferences. This project had to be postponed on account of the difficulties of foreign exchange. The Government of China has undertaken to provide the building materials and technical assistance for this project to serve as a memorial to the late Premier, the Hon. S. W. R. D. Bandaranaike. Chinese architects and engineers are expected to arrive shortly and work will start in the new financial year. It is expected that 3 million rupees will be spent on this project out of an anticipated expenditure of about 17 million rupees".

It is, in the first place, completely shameless for Ceylon to have a foreign country donate a local memorial to a Ceylonese politician, and even though the Coalition Government is very friendly with Red China the whole scheme is an unworthy slight to Mr. Bandaranaike. If this country wants a memorial, in addition to the one at Horagolla, let it build it itself with its own money and its own design and expertise and labour. The Chinese Government would laugh at us if we offered to erect a memorial in Peking to some Chinese national leader. What self respect can Ceylon have if it permits the Red Chinese to erect a memorial here in Colombo, to one of our own Prime Ministers?

There is still time to call the project off because no actual construction work has begun and our suggestion is not merely that Ceylon should not disgrace itself by accepting Chinese charity for a Bandaranaike memorial, but that the memorial (if another is really necessary) should not be an international conference hall even though it is planned to be a sumptuous affair with accommodation for representatives of all the nations of the world. The fact is our Government has done too much "international" talking and too little homework. Whichever government it is that comes to power on March 22nd, it would be an excellent idea for it to make a solemn resolution to go to fewer talking tamashas abroad and to do a little honest work at home. And if another Bandaranaike Memorial is necessary, let it be a hospital or a factory or a school—something, anything, which would be of value to the country and the people, and not just a talking shop.

ලංකා-චීන කතා අරඹී

Will Memorial Hall project have to be abandoned?

OFFICIALS of the Public Works Department who were instructed by their ministry to be ready to go to Peking to discuss the new plans and design for the Bandaranaike Memorial Hall have been waiting for about six months but the trip has so far not materialised.

They are beginning to feel they will never go and the project will be abandoned.

They were first asked to be ready to leave for Peking in February this year. This date was next postponed for June.

Work on the construction of the memorial hall on the Havelock Golf Links, Colombo, has not yet been started because the authorities in Peking are understood to have asked that a delegation of engineers from Ceylon should come there to discuss the new plans and design.

Small hall

Originally a memorial hall costing Rs. 25,000,000 was to be constructed by the Government of the People's Republic of China with accommodation for 540 delegates, 111 observers and 85 representatives. Ceylon was to equip and furnish the hall at a cost of Rs. 3,500,000.

After the National Government assumed office in July 1964, it was decided that a small hall which could accommodate about 60 delegates should be constructed.

Commemorative stamp on May 17

The Ministry of Posts and Telecommunications of Sri Lanka will issue a commemorative stamp of the .15cts. denomination on May 17 (time of issue: 4.15 p.m.) to mark the opening of the Bandaranaike Memorial International Conference Hall.

The stamp, designed by Mr. R. B. Mawilmada, has been printed by Messrs. Thomas De La Rue and Co. Ltd. United Kingdom, by "Delacryl" process, in horizontal format. The colours of the stamp are chalky blue and cobalt, and the size is 45 x 32 mm.

The Philatelic Bureau has, as a special case, made necessary arrangements to make available for purchase at the General Post Office, Colombo, First Day Covers cancelled with the special postmarker that will be used at the Bandaranaike Memorial International Conference Hall Post Office. These covers will be on sale there shortly after 4.15 p.m. on 17th May, 1973.

Official First Day Covers will be made available for purchase at the usual special class Post Offices from 4.00 p.m. to 6.00 p.m. on the day of issue The price of a cover is 25 cts. and a copy of the stamp bulletin will be supplied with each cover.

First Day Covers stamped and cancelled with the special postmarker will be available for purchase from 18th May, 1973 at the Philatelic Bureau, during its working hours.

China's envoy & our PM open BMICH at 3.55 p.m. today

AT 3.55 p.m. today, the Prime Minister, Mrs. Sirimavo Bandaranaike and the Special Envoy of the People's Republic of China, Hsu Hsiang-Chien, will declare open the Bandaranaike Memorial International Conference Hall.

The conference hall, which cost Rs. 35 million to build, is a gift from China in memory of the late Mr. S. W. R. D. Bandaranaike,

Yesterday, the 24 Chinese delegates led by the Special Envoy, arrived by special aircraft at the Ratmalana Airport at 10.50 a.m.

They were met at the airport by the Prime Minister.

After lunch at "Temple Trees", the Special Envoy and party visited the Bandaranaike Memorial Hall in the evening.

This morning, the Special Envoy called on the President at "Janadipathi Mandiraya", the Prime Minister and the Speaker of the National State Assembly, Mr, Stanley Tillekeratne.

He was scheduled to leave for Horagolla at 10.30 a.m. where he is to lay a wreath at the Samadhi of the late Mr, S. W. R. D. Bandaranaike,

斯里兰卡相关报刊剪辑二

Sri Lankan newspaper clippings II

A symbol of friendship between Sri Lanka and China: Chou En-lai

"I believe that the Bandaranaike Memorial International Conference Hall will be recorded in the annals of history as a symbol of the friendship between China and Sri Lanka", the Chinese Premier, Mr. Chou En-lai said yesterday in a message to mark the ceremonial opening of the Hall on Thursday.

Following is the text of his message:

"His Excellency Solomon Bandaranaike the late Prime Minister, was an outstanding statesman of Sri Lanka who devoted all his life to the just struggle of opposing imperialism and colonialism, winning national independence and safeguarding state sovereignty and made positive contributions to the Afro-Asian cause of unity against imperialism. He enthusiastically promoted and developed the friendly relations and co-operation between China and Sri Lanka and was a sincere friend of the Chinese people. The Chinese Government is greatly honoured to have the opportunity of honouring the memory of the late Prime Minister, His Excellency Respected Solomon Bandaranaike, by helping to build the international conference hall.

"The successful completion of the Bandaranaike Memorial International Conference Hall is inseparable from your solicitous concern. You led Government officials of Sri Lanka to the construction site and personally laid bricks for the hall, and repeatedly inspected the construction site — all this was a great encouragement to the Chinese technicians and workers. In the course of the construction of the hall workers and technicians of China and Sri Lanka worked side by side in close coordination and learnt from each other, which has served to deepen the friendship between our two peoples.

"I believe that the Bandaranaike Memorial International Conference Hall will be recorded in the annals of history as a symbol of the friendship between China and Sri Lanka."

Chance to see it later—PM

The Prime Minister yesterday appealed to those who will be unable to witness the ceremonial opening of the Bandaranaike Memorial International Conference Hall to appreciate the problem involved. Arrangements will be made she said to enable the public to see the Hall [illegible]

[illegible] appeal [illegible]

"[illegible] requests [illegible] received from various organisations and members of the public that they be given an opportunity to witness the ceremonial opening of the Bandaranaike Memorial International Conference Hall on May 17, 1973.

"While I would have been extremely happy to have been able to accede to each and every request, I deeply regret that this cannot be done because the Hall cannot accommodate any numbers over its seating capacity. I will be most grateful to all those who would be unable to witness this occasion to appreciate the problem involved. Arrangements will be made for the public to see the Hall on dates to be announced later.

'I would like to make a special appeal to all invitees to the ceremony not to bring along with them any persons other than those indicated in the seating cards sent to them because firstly, such persons will not be accommodated under any circumstances, and, secondly any such action on the part of invitees will create serious dislocation of arrangements already made for the opening ceremony and also cause grave inconvenience to themselves and to those participating in the ceremony.'

Special envoy to open hall

At 3.55 p.m. on Thursday a special envoy from the People's Republic of China Mr. Hsu Hsiang-chien representing Prime Minister Chou En-lai will declare open the Bandaranaike Memorial International Conference Hall — the biggest secretariat complex in this part of Asia.

The imposing structure which stands 94 feet high covering a floor area of 194,000 sq. ft. is an outright gift from China built in memory of the late Prime Minister of Sri Lanka.

One thousand four hundred Chinese and local engineers, technicians and workers went to work on the conference secretariat in November 1970 on a 40-acre block in "Bauddhaloka Mawatha — the old golf links — and completed the job five months before the scheduled date saving five million rupees on the Rs. 35 million project. The saving is computed at Rs. 3 million locally and Rs. 2 million on the Chinese side.

The octagon-shaped main conference hall carries a giant emblem of the Republic of Sri Lanka. The three storeyed building has an assembly hall, a stage and a massive screen in the background, a delegates' lounge over which hang giant chandeliers flown down from China on the first floor. The public gallery and the press lobby are located on the second floor and at the far end of the entrance on the ground floor is a statue of the late Mr. Bandaranaike turned out by a Chinese sculptor. Behind, adorning the wall is a mural depicting an aerial view of the Western province, painted by Chinese artists.

The assembly hall can accommodate 1500 people — 540 delegates, 115 observers, 85 pressmen and members of the public."

(Continued on Page 2).

Bandaranaike Memorial Hall, which is bound to be Colombo's main attraction for Vesak sightseers, looks beautiful and impressive in any light, but it seems to be at its best at night. When this picture was taken, not all the lights were on but the interior lighting and the floodlit fountain give a good impression of what the hall will look like on May 17, when it will be formally opened at about four in the evening. All the lights are due to go on when dusk falls. (Supplement inside)

(Picture by Upali Aturusiri)

Special envoy

(Continued from page 1)

Block II of the building houses offices for delegates consisting of 90 rooms; Block III a committee room complete with a library, telephone exchange, cinema-cum-speech room and air-conditioning plant. The upper floor of this block has two large committee rooms with 430 seats each, two committee rooms of 50 seats each and two more with 30 seats each. Facilities for simultaneous interpretation in seven languages are provided in the large rooms. The Secretariat is located in Block IV with a restaurant, offices and living facilities for visiting delegates.

The internal walls and the columns are finished with marble and the floors with terrazzo and mosaic tiles and parquetry. The ceilings are turned out of plastic coated plywood. Furniture fittings in the building all have been turned out in China.

Most of the Chinese technicians have left. About forty of them remain among whom were till recently a handful of women helping in upholstering. The number will be reduced further till only 16 Chinese technicians will be kept for few months here until the Ceylonese staff at the Memorial Hall are fully acquainted with the maintenance system.

All arrangements for the ceremonial opening on Thursday, May 17 are being made by the Ministry of Defence and Foreign Affairs

HALL OPENING—HUGE CROWDS EXPECTED

Police are catering for huge crowds, expected at the Ceremonial opening of the Bandaranaike Memorial International Conference Hall on Thursday, May 17.

A closed area will be declared from 2-30 p.m. till 7-30 p.m. between Bullers Road (from SLBC Junction) to Kanatte roundabout, Wijerama Road to Gregory's Road and Gregory's Avenue. This area will be out of bounds for vehicles other than those in which special invitees will be travelling in.

Vehicles carrying refreshments will be permitted into the area.

There will be four days of official ceremonies in connection with the opening of the Bandaranaike Memorial International Conference Hall. The ceremonies will begin with the opening of the Hall at 3-55 p.m. by a dignitary of the Peoples Republic of China followed by a public meeting.

May 17 is a Public Holiday being an additional Holiday for Vesak.

The CTB will operate special bus services to cater to the heavy rush of people.

The Colombo Municipal Commissioner, Mr. B. A. Jayasinghe told the "Observer" that the CMC will keep the closed area clean. Drinking water and toilet facilities will be provided within the vicinity on May 17.

On May 18 there will be a Cultural Show at the BMICH the highlight of which will be the ballet "Pahanin Pahana." On this day the Police will declare the same area closed from 5.00 p.m. to 9.00 p.m.

There will be all-night Pirith on Saturday, followed by a Sanghika Dana.

THE CEYLON DAILY NEWS, TUESDAY, MAY 15, 1973 — FIVE

'Historic event'

WILLIAM GOPALLAWA, President.

'An act of homage'

SIRIMA R. D. BANDARANAIKE, Prime Minister.

'A symbol of friendship'

HUANG MING-TA, Ambassador Extraordinary and Plenipotentiary of the People's Republic of China to the Republic of Sri Lanka.

'Ahead of schedule'

PIETER KEUNEMAN, Minister of Housing and Construction.

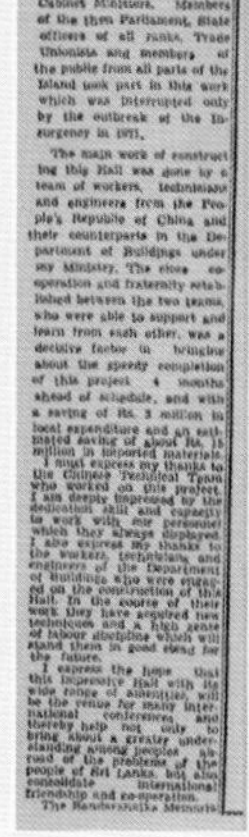

සමරු හලේ අනුරුව නිමැවූ කලාකරු

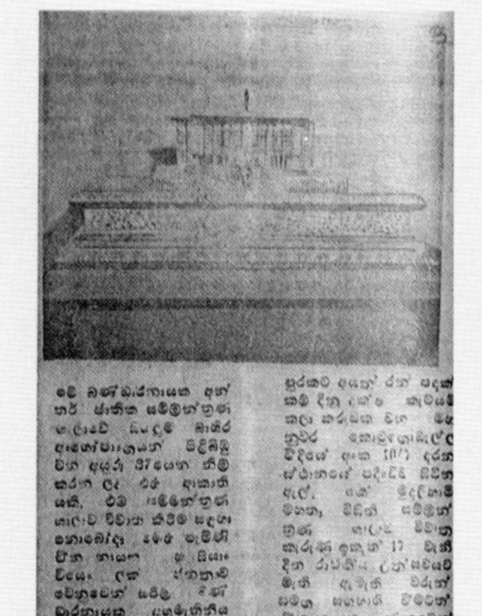

CHINESE EXPERTS HERE

Memorial Hall: plans are drawn

(By an "Observer" reporter)

PRELIMINARY work in connection with the building of Ceylon's biggest Conference Hall—The Bandaranaike Memorial Hall— has started with the arrival of three Chinese building experts.

This team is now drawing up plans and specifications for the Hall in consultation with the Ceylon Government.

The Hall will be built by the Government of People's China and presented to Ceylon as a tribute to the memory of the late Prime Minister, Mr. S. W. R. D. Bandaranaike. It will be sited at the Golf Links in Bullers Road.

When the Prime Minister of China, Mr. Chou En-Lai came here early this year, he evinced interest in a government scheme to build an International Conference Hall and offered to build it entirely at the expense of the Chinese Government.

According to reliable sources, building operations are expected to start by early next year. The Chinese experts have already surveyed the site for the Hall and approved it.

斯里兰卡报刊对工程组组长郝辅堂（左）和副组长兼总工程师由宝贤（右）的报道

A Sri Lankan newspaper report on team leader Hao Futang (left) and his deputy and chief engineer You Baoxian (right)

NEWSPAPER CUTTING FROM

...nd the 'Great ...na'

...oncept of labour that phy...cal exercise was an essen...ial prerequisite to human ...eings, whatever their call...ug in life be.

No little wonder then that ...e Chinese team of techno...gists and workers consisted ... a Lady Doctor, an accoun...ant and a few others en...aged in other professions. ...hough this lady doctor was ...signed the task of looking ...fter the health of the 460 ...hinese engaged on the project, she was also engaged in the construction pushing the wheel barrow up and down the site in the sweltering heat!

12 SUMMERS

Referring to the heat, Mr. Tang jokingly remarked that during his short stay in Sri Lanka he had experienced 12 "summers", meaning that unlike in China where the seasons were distinctly felt, they experienced the sweltering heat all throughout the year!

"Even your Prime Minister Mrs. Sirimavo Bandaranaike was very much concerned with the expeditious completion of the building and our men worked with much enthusiasm rubbing shoulders along with their Sri Lanka counterparts to see that this 'Friendship Hall' was completed before schedule so that our ties could be further strengthened. The speed with which the building was completed demonstrated the speed with which we wanted to consolidate the existing friendship between the two countries", the two engineers added.

Commenting on the building itself they told me that this was the first one they had constructed in Asia though they had also constructed a similar structure in Guinea and undertaken the construction of another Conference hall in Sudan.

Mr. Hoa Fu-Tang, team leader and his Deputy Mr. Yu Pao-Hsien.

The Bandaranaike Memorial Conference Hall will be the biggest Conference Hall in the entire East, they added, though the Hall of the Peoples Republic of China in Peking had seating accommodation for 10,000.

The designing for the Hall commenced as far back as 1964 with teams of Chinese and Sri Lankan architects and engineers working out the details and though the construction took place six years later, only a few modifications had been made in the original design. For instance the stage in the hall was never envisaged in the original design though it was incorporated at a later date.

The workforce was divided into three main categories: Administrative, designing and construction. The latter category of workers was subdivided into various groups dealing with the structure, decoration, carpentry work, and equipment installations. These were further divided into groups of ten men each under a leader. Every view put forward by these smaller groups were summarised and later put to the Sri Lanka personnel for their approval before any suggestions or views were accepted as final.

They added that there were instances when special tools had to be used by the Chinese team. These tools had been turned out for greater output and the Ceylonese workers readily picked up their use and even the new techniques adopted by the Chinese team.

The two Engineers who spoke to me through their Interpreter Mr. Ku Yung-Ching had high praise for the age-old traditional hospitality of the people of this country and their generosity. "We hope and look forward to seeing your country flourishing under the able leadership and guidance of your Prime Minister Mrs. Sirimavo Bandaranaike. Our love and esteem for your country and your people shall remain ever green in our memories however far we may be from you all", they said.

斯方报纸还对施工和建设人员进行了报道，名为《中国（援建）大厦背后的人》，作者为P.Balasuriya。

文章中写道：

“班厦”现在完工了，这样一个壮丽和令人感动的工程仅用了30个月就完成了，并且这个建筑将会给这个国家带来国际荣耀。至于这个巨大的工程提前工期一个月完工，在采访中国技术专家时，中国专家讲：“这很简单，这是团队精神，双方合作的结果”。并还风趣地说：“由于天气炎热，在斯里兰卡短暂居住的3年里，我们经历了12个夏天”。就是说，与中国的气候四季分明不同，他们全年都在过夏季。

A Sri Lankan newspaper reported on construction personnel with the headline, "The People behind the China-aided Building", written by P. Balasuriya.

P. Balasuriya wrote:

"BMICH, a magnificent and moving project, only took 30 months to complete, and this building will bring Sri Lanka international glory. While asking the Chinese technical expert why such a large project could be completed a month in advance, the Chinese expert said, 'This is very simple, it is a result of team work and cooperation.' Referring to the heat, the expert jokingly remarked that during his short stay in Sri Lanka he had experienced 12 'summers', meaning that unlike in China where the seasons were distinctly felt, they experienced the sweltering heat all throughout the year!"

美丽花絮
Beautiful Highlights

1973年“班厦”落成不久，开了几次国际会议；与会代表对大厦的外貌和内部装修都认为非常美观，使用功能也非常好，都称赞不已。中共中央办公厅和国务院办公厅基于上述反映，通报表扬“班厦”及其建设者。

1974年，“班厦”的主要设计者和建设者戴念慈总建筑师和由宝贤总工程师，被特别邀请以科技人士身份，参加了首都庆祝“五一”国际劳动节北京颐和园地区的庆祝活动，并刊登于人民日报。晚上还受邀在天安门观礼台观看了“五一”烟火，璀璨的烟火在天安门广场上的夜空华丽绽放，给人们留下美好难忘的记忆。

Shortly after the completion of BMICH in 1973, it hosted several international conferences. Delegates all gave much praise, saying the building looks beautiful from the outside appearance to the interior decoration, and it is very functional. Because of the warm reactions, the General Office of CCCPC and General Office of the State Council circulated a notice of commendation of BMICH and its construction team.

In 1974, chief architect Dai Nianci, main designers and engineers of BMICH and chief engineer You Baoxian were invited as technical professionals to participate in the May Day celebration in Beijing's Summer Palace, and this event was published in the People's Daily. In the evening, they were invited to watch the May Day fireworks at the Tiananmen Reviewing Stand, the dazzling fireworks exploding in the night sky, leaving nice and unforgettable memories.

1974年5月1日国际劳动节在北京颐和园举行盛大的庆祝会
Grand May Day celebration held in Beijing's Summer Palace on May 1, 1974

第 二 章
“班厦”工程技术
Chapter Two Engineering Technology of BMICH

项目的确定
Project Identification

一　接受设计任务

1964年协议签订后，中国外经部决定由当时的建筑工程部北京工业建筑设计院（现中国建筑设计研究院）承担设计任务，并下达了周总理对设计工作提出的三项原则。

周总理指示：“班厦”是中国援外项目中第一个大型文化建筑项目，同时，也是锡兰（现斯里兰卡）未来举行重要活动和召开国际会议的场所，“纪念班达拉奈克国际会议大厦”具有非比寻常的重要性。为此特别提出了设计工作的三项原则是：

- 规模应适合锡方的国情和要求，不宜过大；
- 建筑要适应热带自然条件和要求，造型要尽量体现当地风格；
- 建筑内部设施应尽量采用先进技术和装备。

(I) Accepting the Design Task of BMICH

After signing the agreement in 1964, the Ministry of Foreign Trade and Economic Co-operation assigned Beijing Industrial Building Design Institute of the Ministry of Works (now China Architecture Design & Research Group) to design Bandaranaike Memorial International Conference Hall (BMICH), and issued three design principles proposed by then Premier Zhou Enlai.

Premier Zhou Enlai stated that BMICH is the first large-scale cultural project aided by China, and it will host important events and international conferences in Ceylon (now Sri Lanka). Due to its significance, the three design principles put forward for the project were:

- The building scale shall be appropriate to fit Sri Lanka's conditions and requirements;
- The building shall adapt to the tropical climate and requirements, and the building shape shall reflect the local style;
- The internal facilities and equipment shall adopt advanced technology.

考察组专家在使馆办公楼前的合影。左起：刘茂堂、杨芸、袁镜身、由宝贤、韩鸿钧、范世凯。右起：马贞勇、鲍兆华、戴念慈

A group photo of delegation experts in front of the Embassy. From left: Liu Maotang, Yang Yun, Yuan Jingshen, You Baoxian, Han Hongjun, Fan Shikai From right: Ma Zhenyong, Bao Zhaohua, Dai Nianci

在接受“三项原则”的同时，北京工业建筑设计院还收到外经部转交的一份设计方案。这个方案是其他国家的建筑师为锡方做的设计，外观好似两个贝壳合在一起，又像中国的烧炭形火锅。对这个方案，领导们指出：一是不美观、不实用、不经济；二是我国经援的建筑，应该用我们的设计方案。

设计院当即成立考察组，由袁镜身副院长任组长，成员包括总建筑师戴念慈、主任建筑师杨芸、结构工程师由宝贤、给水排水工程师韩鸿钧、空调工程师刘茂堂、电气工程师范世凯、模型技师鲍兆华、翻译马贞勇等，于1964年6月赴斯里兰卡考察。

二　现场考察

到了科伦坡，设计考察组受到了当地政府官员的热情接待。大使馆将考察组的工作生活安排照顾得非常周到。

为了“班厦”的设计，考察组在斯里兰卡历时4个月，考察了大半个岛屿，对斯里兰卡的历史、人文习俗、天文地理、建筑的民族形式都进行了考察，并有了更深刻的理解，对科伦坡的自然风景、建筑风格等也有了进一步的体会。

In addition to the Three Principles instruction, Beijing Industrial Building Design Institute received a design scheme from the Ministry of Foreign Trade and Economic Co-operation. It was submitted by architects from other countries for the Sri Lanka project, and featured the building looking like two seashells combined together, or a Chinese charcoal burning hotpot. Chinese design leaders believed the scheme was not beautiful, practical or economical, and it was decided that China should adopt its own design for this China-aided project.

Beijing Industrial Building Design Institute set up an inspection delegation to visit Sri Lanka in June 1964. The delegation was led by vice president Yuan Jingshen, and its members included chief architect Dai Nianci, principal architect Yang Yun, structural engineer You Baoxian, plumbing engineer Han Hongjun, air conditioning engineer Liu Maotang, electrical engineer Fan Shikai, model technician Bao Zhaohua and interpreter Ma Zhenyong.

(II) Site Inspection

Colombo government officials warmly received the design delegation, and the Embassy made good arrangements of the schedule for the touring group.

The delegation stayed in Sri Lanka for four months, during which, they walked around more than half the island and learned more about the country's history, cultural customs, astronomy, geography and national ethnic design in Sri Lanka, had further understanding of the natural landscape and architectural style of Colombo.

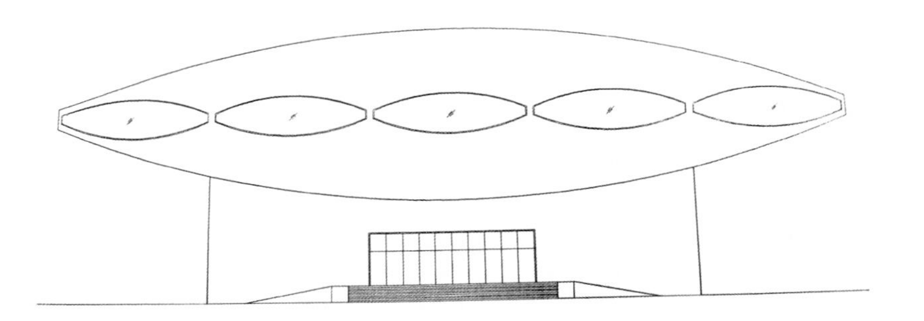

形似火锅的方案示意图
Hotpot-shaped building diagram

这里热带植物繁茂，由于曾受殖民统治，具有斯里兰卡民族特色的建筑与具有欧洲风格的建筑在科伦坡市区相互交错，形成了独具特色的建筑景象，让人在感慨的同时不禁又有新的感悟。

The Sri Lankan capital is home to lush tropical plants. Due to its history of colonial rule, buildings with Sri Lanka's national characteristics and European-style buildings coexist, thus forming an unique architectural landscape and giving people a new sentiment while having emotional attachments to them.

独立广场的独立纪念堂柱子
Columns of Independence Memorial Hall on Independence Square

首先，考察了拟建“班厦”的原址，那原是一个高尔夫球场，有一栋原为俱乐部用的简易二层小楼，后作为工地用房。由于荒废多时，周围杂草丛生。

阿努拉达普拉、坡伦那鲁、康提是斯里兰卡的三座主要古城，至今保留着不少古塔和古代皇宫遗址。

阿努拉达普拉城最著名的是有着1200年历史的罗凡威里萨佛寺，寺中白塔高50多米，直径70多米，塔基座的四周雕刻着数百头大象。虽经风雨沧桑已有些风化，但昔日巍然屹立的雄姿风采，仍依稀可见。

The delegation firstly inspected the former site of BMICH, which was on a golf course, and it occupied a simple two-storey building originally used as a clubhouse which was later a site office. The site was overgrown with weeds since it was abandoned for a long time.

Anuradhapura, Polonnaruwa and Kandy are three major ancient towns of Sri Lanka, where still stand many well-preserved ruins of ancient towers and royal palaces.

The Sacred City of Anuradhapura is most famous for the Ruwanwelisaya Stupa of more than 1200 years of history. The white stupa is 50 meters high and its diameter reaches 70 meters. Surrounding the stupa base are carved with elephants. Although the stupa is slightly weathered through years of vicissitudes, the stupa still maintained its original ancient style.

科伦坡保留着许多用花岗石砌成的古代佛塔、台基和塔身。那里的佛塔主要有细而高的尖形塔和矮而粗的圆形塔。市中心有一座保留得很好的古庙，大殿门口雕刻精致的石狮子，抬头远望，注视着一切，好似威震这方领地。古庙里还有一座宛如被一枝莲花托起的塔，看上去十分美观优雅。还有的塔宛如雅致的阁楼。在科伦坡沿印度洋海岸一带，有一座1837年建的钟楼，由港口灯塔改建而成，登上钟楼，可鸟瞰科伦坡全景，美不胜收。

坡伦那鲁城给人印象最深的是四座形态庄严的大佛，由一整块小山坡状的花岗岩雕刻而成，在周围参天古树的衬托下，更显这座城市的古朴和沧桑。那里的野猴不怕人，在树上跳来跳去，当有人下车时，它们就跳过来向人们摇头摆手要吃的，就像我国现在野生动物园里的场景。

Colombo is home to ancient granite pagodas, stylobates and towers. Pagodas there are generally divided into two types, thin and tall towers, and thick and short round towers. An ancient temple is located in the downtown area, and the main temple hall has exquisitely carved stone lions looking up in awe of the place. There is a tower that appears to be supported by lotus in the ancient temple, which is beautiful and elegant. Other towers look like elegant attics, with overhanging eaves and brackets, and are lifelike. Along the coastal line of the Indian Ocean, there is a bell tower that was rebuilt from a port light tower in 1837. By climbing up the tower, one can have a bird's eye view of Colombo's beautiful scenery.

The ancient city of Polonnaruwa is most impressive with the four Buddha statues, which were sculptured with a whole piece of granite. Against high ancient trees, the city appears more primitive. Wild monkeys there seem unafraid of humans, jumping in the trees freely or begging for food from passersby, similar to wildlife parks in China.

1. 考察时，在不同建筑风格相互交错的街景上的留影
   A photo taken on a street with buildings of different architectural styles during the site inspection
2. "班厦"原址面貌，小楼后来作为工地用房
   The former site of BMICH, where the small building was later used as a site office
3. 大厦竣工前检查组在庙宇白塔前留影，左起：斯方陪同人员、李子武、郝辅堂、乔林、由宝贤、斯方司机、王挺、检查组成员
   The inspection delegation in front of White Tower
   From left: Sri Lankan personnel, Li Ziwu, Hao Futang, Qiao Lin, You Baoxian, a Sri Lankan driver, Wang Ting and an inspection team member
4. 斯里兰卡阿努拉达普拉城
   The City of Anuradhapura in Sri Lanka

4

康提城坐落在斯里兰卡中部山区，是全国第二大城，也是锡兰王朝最后的古都。这座景色宜人的山城里保留着斯里兰卡最为神圣，建筑最为宏伟的佛教寺庙——佛牙寺。

Situated in the central mountainous area of Sri Lanka, Kandy is the second-largest city and the last ancient capital of the Ceylon Dynasty. In this city with beautiful scenery, is the most sacred and magnificent Buddhist temple -- the Temple of the Tooth.

佛牙寺始建于15 世纪，周围有清泉环绕，景色极为清雅幽静。由于这里四季温暖，房屋都十分通透，建筑材料主要为木质，装饰以精美彩绘和木雕为主。那雕花装饰精美得仿佛能闻到花香，更散发着斯里兰卡那独具特色、充满佛教文化的民族文化气氛。

Built in the 15th century, the Temple of the Tooth is surrounded by crystal-clear spring water, and the atmosphere is elegant and quiet. The climate here is warm all year round, so the buildings are transparent and mainly made of wood, with exquisitely painted and carved wood ornaments. The engraved flower ornaments are so enchanting, it is as if one could smell the flowers' fragrance, reflecting the unique national cultural atmosphere full of Buddhist cultural characteristics.

康提佛牙寺
Temple of the Tooth in Kandy

在斯里兰卡，佛牙的地位堪比王位。寺内还有一幅浮雕，讲述了一段关于佛牙来历的故事。公元371年，古印度国王在战乱中，命女儿将佛牙藏在发髻里，送往师子国供奉。彩绘浅浮雕将公主的发髻上方雕成放射的光芒，表示佛牙藏在那里。

In Sri Lanka, Buddha's tooth has a position that is comparable to the throne. The relief carvings in the temple tell the story about the origin of Buddha's tooth. In AD 371, a war broke out, and the king of ancient India ordered his daughter to hide Buddha's tooth in her hair ornament and sent her to what is now Sri Lanka. On the colored relief carving, the top of the princess's hair ornament emits light, indicating the tooth relic is hidden there.

1. 由宝贤在考察时
A photo of You Baoxian conducting the onsite inspection

2. 佛牙寺的彩绘浅浮雕
Colored relief carving in the Temple of the Tooth

3. 珍藏佛牙的八角金塔
Gold octagonal pagoda where Buddha's tooth is kept

考察组的部分成员在住所大门前的合影。左起：鲍兆华、戴念慈、袁镜身、由宝贤、马贞勇

A group photo of some delegation members at the residence gate.From left: Bao Zhaohua, Dai Nianci, Yuan Jingshen, You Baoxian, Ma Zhenyong

佛牙寺的山坡下有个八角亭，保存着有一千多年历史的“贝叶经”（贝叶经是将佛经写在贝多罗树叶上的经书）。“班厦”竣工，中国徐向前特使来访时，斯方就是把敬养佛牙的金塔打开放在这个八角亭里请特使瞻仰的。金塔打开，内套有7座小金塔，还有玻璃罩，塔里有一朵金莲花，花心有一玉环，直径约5厘米。佛牙就放在玉环上。

通过对斯里兰卡建筑的考察，考察组认识到，在斯里兰卡这个以佛教为主为尊的国家，有以八边形建筑为世间至尊的建筑认同，尤其是象征着国王权势的佛牙寺建筑，对后来的方案设计有了较大的启发。同时我们也了解到，斯里兰卡的建筑大多数都是木质结构，当地的木材很好，都是硬木，质地很坚硬。水泥、砂子、石子的质量都不错。

In an octagonal pavilion on the hillside of the Temple of the Tooth, the "Palm-leaf Manuscript" (manuscript written on leaves) has collected over one thousand years of history. Upon completion of BMICH, Chinese special envoy Xu Xiangqian visited Sri Lanka, and the Sri Lankan side placed Buddha’s tooth in the gold tower in the octagonal pavilion and invited the special envoy to see it. The gold tower has seven small gold towers in it and a glass shield. Inside it is the center with a gold lotus with a jade ring (about 5cm in diameter).

Based on their investigation of Sri Lankan architecture, the delegation realized the country honors the Buddhism, and the architectural sector has a common view that octagonal buildings are supreme, especially the Temple of the Tooth that symbolizes the king’s power, which greatly inspired later designs. Meanwhile, they also learned the majority of Sri Lankan buildings are wooden structures, and Sri Lanka has good hard wood material, as well as cement, sand and stone of high quality.

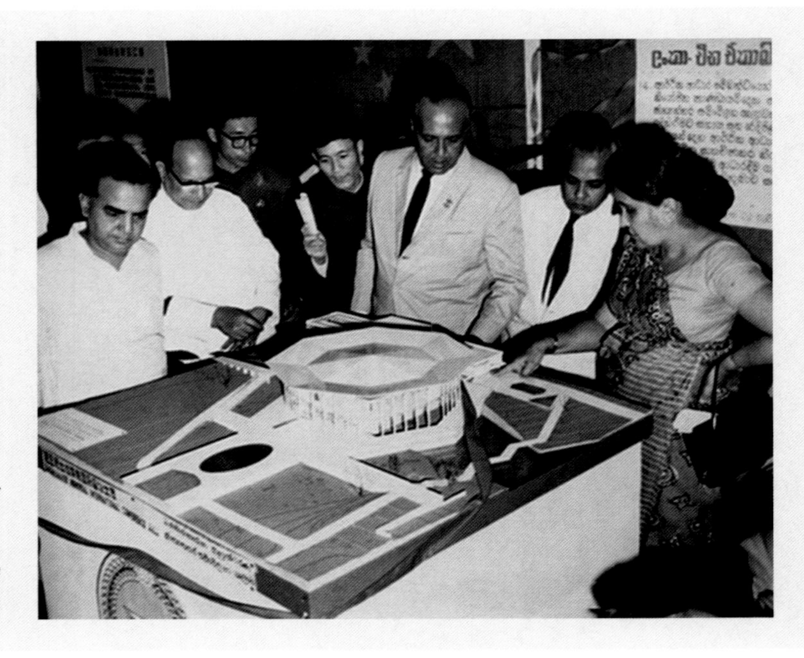

总理班夫人正在审查“班厦”模型

Prime Minister Sirimavo Bandaranaike inspects the BMICH model

三　收集资料

在考察中，收集了大量有关水、暖、电等各个设计方面的基础资料，为“班厦”的方案设计工作做了充分的准备。对他们的建筑风格、建筑材料、气候变化、地理环境做了细致的分析。建筑师千方百计做好这个建筑，既要比国外方案美观，又要有斯里兰卡的民族特色，还要有新时代的气息；要做出外貌新颖，功能实用，设备先进，技术合理安全，并且能让总理班夫人和其他高级官员接受的方案。

四　初步确定

考察组在科伦坡夜以继日地工作了3个月，完成了八角形外柱廊式的方案。方案初步确定后，模型技师鲍兆华按照这个方案，做了一个1：200比例的精致模型。模型是用塑料做的，把大厦的八角形自下而上，层层向外悬挑，把那种腾空出世的外观表现得淋漓尽致。庭院、喷水池、草地、停车场、树木和灯光效果都表达了出来，连大厦室内也能亮灯。这样的模型制作，在当时的技术和材料条件下，可以说已经达到了登峰造极的程度。

(III) Information Collection

The delegation collected a large amount of basic data for water, heating, electrical designs, to prepare for the schematic design of BMICH. They also conducted careful analysis of the architectural style, building materials, climate changes and geographical environment. By incorporating both the beautiful appearance more attractive than the foreign design scheme and national features of Sri Lanka as well as modern touches, the architects made every endeavor to design a building with a new look, practical functions, advanced equipment, rational and safety technology, and a scheme that satisfied Prime Minister Sirimavo Bandaranaike and other Sri Lankan senior officials.

(IV) Preliminary Determination

The delegation worked in Colombo for three months day and night and finally completed the external octagonal colonnade design. After it was initially confirmed, model technician Bao Zhaohua made a 1：200 model according to the design. The plastic model incisively and vividly reflected the splendid appearance of the building, with its design of octagonal overhanging layer by layer from the bottom upwards. The model also presented such elements as the courtyard, fountain, lawn, car park, trees and lighting effect, and even indoor lighting. It is safe to say this model reached the highest level of refinement under the technical conditions and materials available at the time.

1964年8月下旬的一天，总理班夫人接见了中国驻斯里兰卡大使谢克西与袁镜身、戴念慈、杨芸、由宝贤等考察组成员。考察组将精致的八角形大厦模型送给总理班夫人，总理班夫人仔细地观看了模型之后，她兴奋地说："感谢周恩来总理赠送给我们的最高礼物，感谢中国专家做出的卓越贡献。这座大厦就是锡中友谊的象征！"

10月中旬，考察组从科伦坡回国后，组织了50多人参与"班厦"的全部设计工作,其中有天津大学工民建系应届毕业的17名实习生。

"班厦"新颖独特的造型设计为建筑的结构设计带来了巨大挑战。

One day in late August 1964, Prime Minister Sirimavo Bandaranaike met with Chinese ambassador to Sri Lanka Xie Kexi and other delegation members, including Yuan Jingshen, Dai Nianci, Yang Yun and You Baoxian. The delegation sent an exquisite octagonal model of BMICH to Prime Minister Sirimavo Bandaranaike who, after inspecting it carefully, excitedly said, "Thank Premier Zhou Enlai for sending us the highest gift, and thank the outstanding contributions the Chinese experts have made for the project. BMICH is the symbol of China-Sri Lankan friendship!"

In mid-October of 1964, after returning to China, the delegation organized 50 personnel to work on BMICH, including 17 interns newly graduated from the Civil Engineering Department of Tianjin University.

The unique and distinctive shape of BMICH presented a great challenge for the project's structural design.

精致的"班厦"模型

Refined model of BMICH

“班厦”工程技术
Engineering Technology of BMICH

“班厦”立面图片
BMICH façade

一　建筑设计

总建筑师戴念慈经过这一段时间的酝酿和深思熟虑，设计了一个新颖的古典构图、现代主义手法、有许多独到之处的方案。

“班厦”是一座纪念性建筑，这个纪念性文化建筑，应体现出它悠久美好的传统文化和走向独立后，有着光明美好的现代感。戴念慈总建筑师通过考察，对其自然环境和人文环境有了较深刻的理解，提出在建筑设计中采用：对称的八角形布局及柱廊式的经典做法。这种形式端庄、稳定、纪念性强，能很好地表现“班厦”的特性。

(I) Architectural design

After long-term preparation and deliberation, chief architect Dai Nianci presented a design that combined novel classical composition, modernist techniques and distinctive features.

As a memorial cultural building, BMICH should reflect the long and wonderful traditional culture and move towards bright and beautiful modern sense after Sri Lanka's independence. After the investigation, chief architect Dai Nianci had a deep understanding of Sri Lanka's natural and human environment, and he proposed the classical practice of a symmetrical octagonal layout and colonnade should be adopted in the architectural design, as this style was elegant, stable and highly commemorative, and can best demonstrate the characteristics of BMICH.

1. “班厦”总括

“班厦”建筑分为主体建筑及附属建筑，由四个段组成。总占地13公顷，总建筑面积32540平方米。

1. BMICH overview

BMICH is comprised of a main building and auxiliary buildings, and has four blocks. The project covers an area of 13 hectares and a total floor area of 32,540 square meters.

“班夏”总平面图

BMICH general layout plan

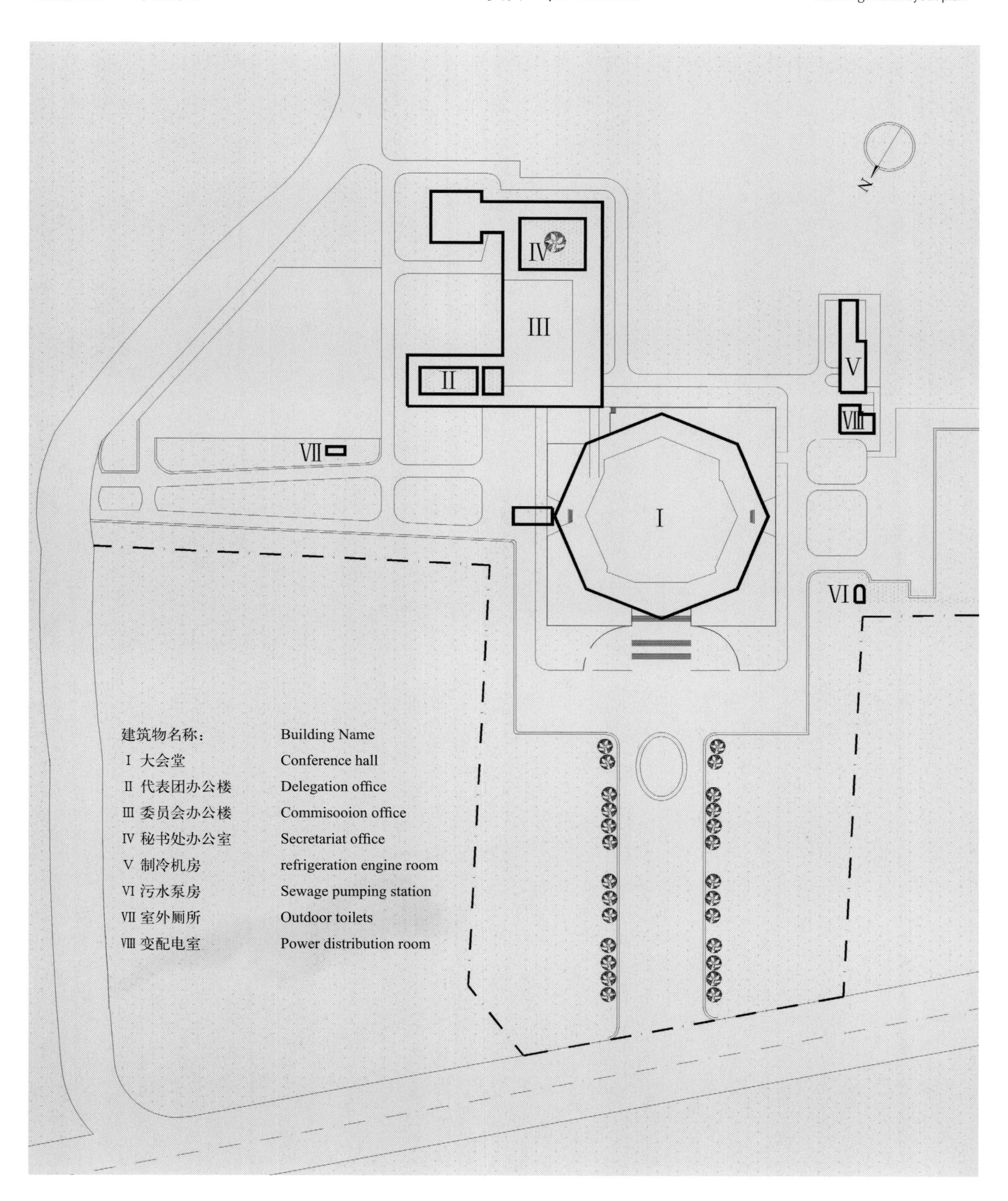

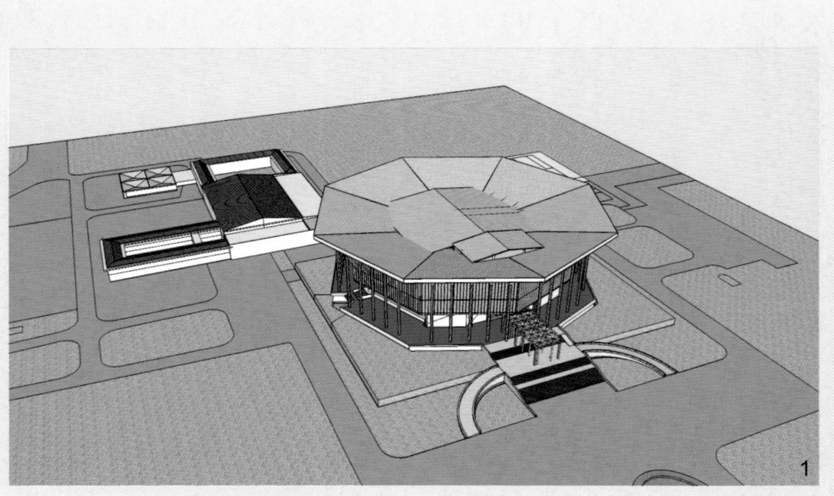

1.“班厦”模型
BMICH model

2. 鸟瞰“班厦”
A bird's view of BMICH

“班厦”大院共有四个大门。北门为正门，对应大厦的主入口；其他均为次入口。次入口处设有停车场和绿地草坪，方便活动时停车。

正门旁矗立着由僧伽罗、泰米尔、英、中四种语言写就的“纪念班达拉奈克国际会议大厦”名称标牌。

BMICH has four gates, where the north gate is the front one corresponding to the main entrance, while the other gates are secondary entrances where the parking lot and lawn are available for parking convenience during events.

Next to the front gate is a sign that says "Bandaranaike Memorial International Conference Hall" written in Sinhala, Tamil, English and Chinese.

1. “班厦”正门
Main entrance of BMICH

2. 正门旁矗立着由僧伽罗、泰米尔、英、中四种语言写就的“纪念班达拉奈克国际会议大厦”名称标牌。
The sign saying "Bandaranaike Memorial International Conference Hall" written in Sinhala, Tamil, English and Chinese next to the front gate

主入口有喷水池及绿化、90根高耸的旗杆
At the main entrance there is the fountain, greenery and 90 towering flagpoles

主入口喷水池及绿化。大佛右侧的建筑是中国驻斯里兰卡大使馆
The fountain and green area at the main entrance, the building on the right side of the giant Buddha is the Chinese Embassy in Sri Lanka

2. 主体建筑（即大会堂）

斯里兰卡是一个热带岛国，气候炎热，林木茂密。为适应这种自然环境，“班厦”主体建筑使用了热带建筑惯用的，也是斯里兰卡人民喜欢的颜色——白色，以减少对太阳辐射热的吸收。同时，使用大挑檐、柱廊、通透的竖向遮阳板墙，有利于建筑的通风降温。八角形的白色建筑在一片葱绿的衬托下，更显轻盈开朗的热带建筑特色。

主体建筑是一段。为柱廊式对称八角形的大会堂。建筑高30米；大会议厅设于标高8米楼板处；楼板平台向外挑出，挑檐最长处达7米，主楼屋面再挑出12米屋檐。大厦自下往上，一层一层向外挑出，看上去有一种腾空出世之势。

2. Main building (Assembly Hall)

As a tropical island, Sri Lanka has a hot climate and lush forests. To adapt to such a natural environment, the main building of BMICH is painted white, a common color for tropical architecture and a color favored by Sri Lankan people, as white can reduce the absorption of solar radiant heat. Meanwhile, transparent vertical sun-shade plates on the walls with large overhang and colonnade facilitates building ventilation and cooling. The white octagonal building against the background of light green highlights the light and open tropical architectural features.

Block I is the main building, with a 30m-high symmetrical octagonal and colonnade-type assembly hall; the large assembly hall is set on a floor with an elevation of 8m; floor platform has an overhang, with the longest eave at 7m, and the main building roof overhangs by another 12m. The building overhangs from the bottom up and layer by layer, reflects the splendid appearance of the building.

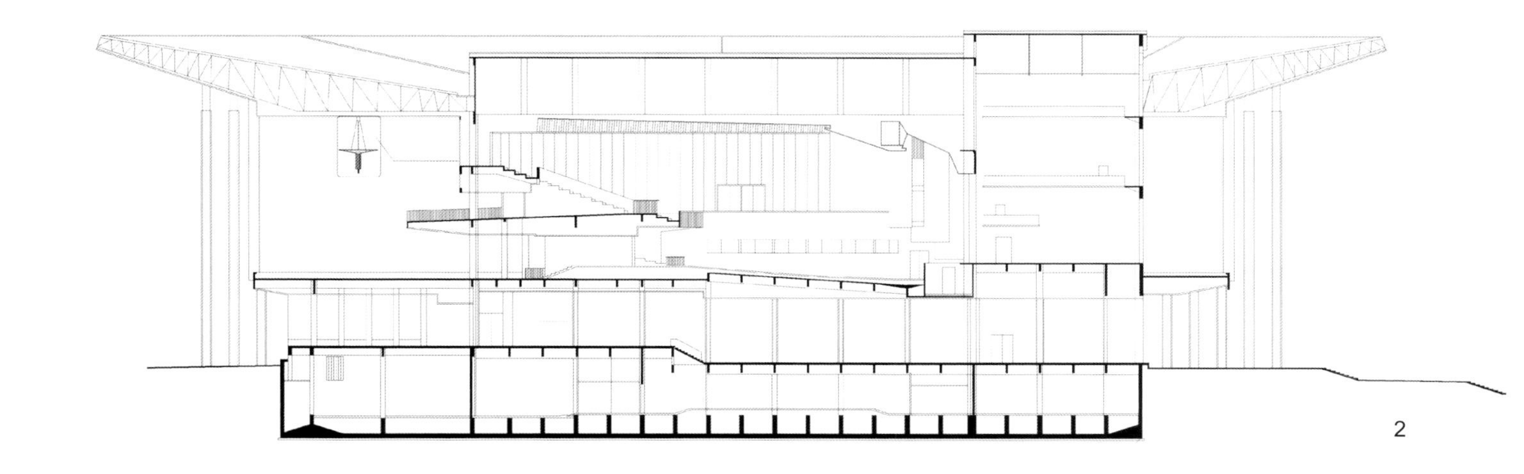

1. 主体建筑（大会堂）夜景璀璨
The main BMICH building (Assembly Hall) lit up at night

2. 大会堂（一段）剖面图
Section of Block I

1. 国徽高高悬挂，大挑檐和金色柱头在蓝天下的壮美情景
The national emblem hanging high, large overhanging eaves and a gold chapiter under blue sky

2. “班厦”水彩画（戴念慈绘）
Watercolor painting of BMICH (by Dai Nianci)

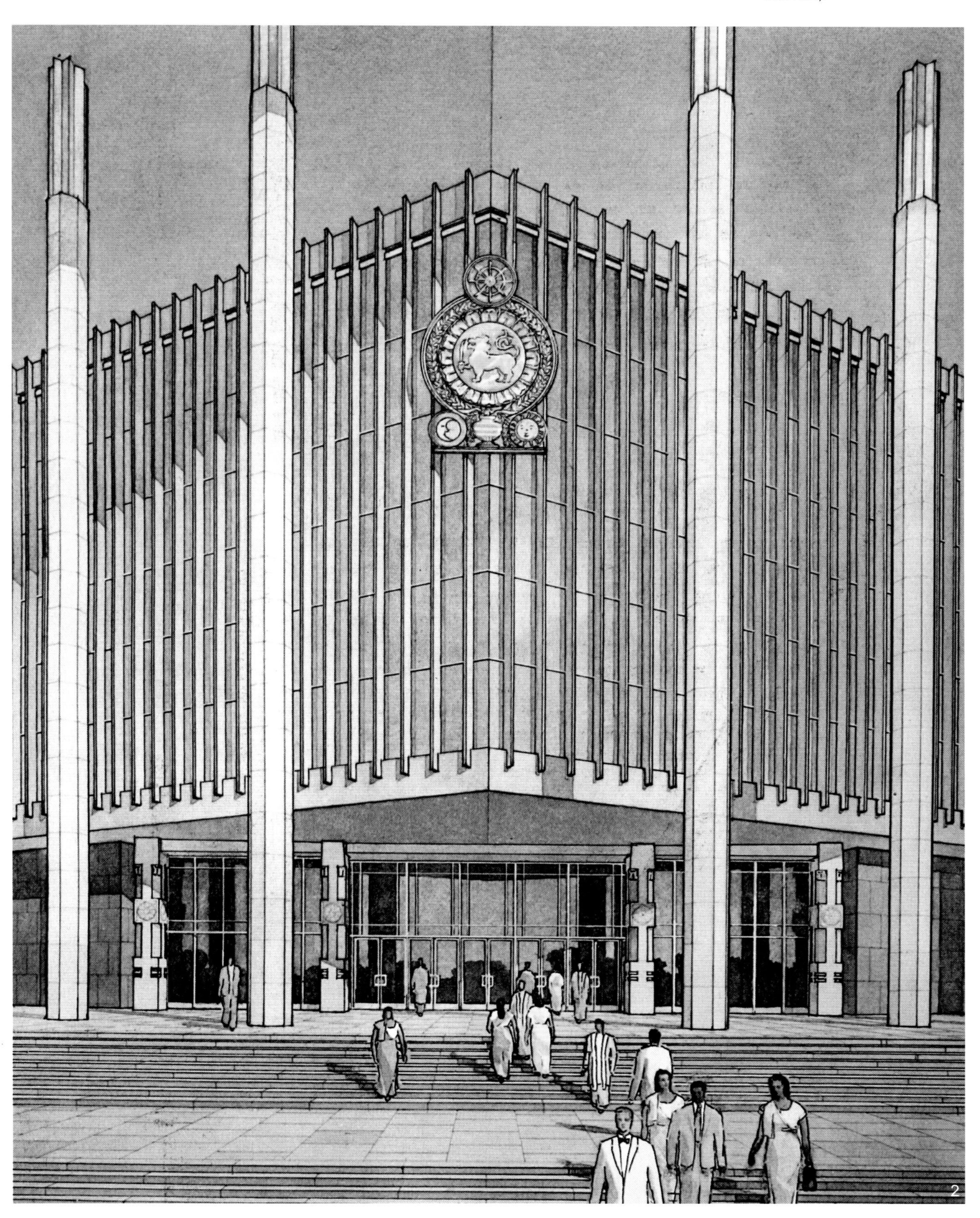
2

1. 四根具有斯里兰卡民族风格的汉白玉柱子
   Four white marble pillars in the Sri Lankan national style
2. 正门两边的红色大理石墙
   Red marble walls on both sides of the main entrance
3. 八角形大会堂其中一面的5根外廊柱
   Five colonnades on one face of the octagonal assembly hall

以狮子图案为中心的国徽悬挂于建筑正门上方。既突出了这个文化建筑的庄严性，也突出了狮子象征的这个国家刚强、勇敢的悠久文化。

正门两侧是四根具有斯里兰卡民族风格的、精致的汉白玉雕花柱。两侧墙壁则使用的是斯里兰卡人民喜欢的红色大理石墙面。

The national emblem with a lion in the center is hung above the building's main entrance, reflecting the solemnity of this cultural building and symbolizes the national culture of strength and braveness, just like a lion.

On both sides of the main entrance, there are four exquisite white marble-carved columns in the Sri Lankan national style, and the walls on both sides are in red marble, which is favored by the Sri Lankan people.

八角形大会堂的外廊柱，每面5根，共40根纤细挺拔俊美的柱子，组成高达24米的外廊，那每根纤细俊美挺拔的柱子与已故总理班达拉奈克先生墓地周围五根巍然屹立象征和平共处五项原则的5根柱子遥相呼应。

外廊柱使用我国山东出产的雪花白大理石贴面，意为把中国凉爽的雪花带到美丽的斯里兰卡，带到庄严俊美的“班厦”。每根柱子均配有尽显尊贵的金色柱头。金色柱头承托宽大的挑檐，在阳光下永远熠熠生辉。

The 24m-high verandah in the octagonal assembly hall is composed of 40 slender, tall and handsome colonnades, surrounded by five columns. These colonnades correspond to the five columns representing the Five Principles of Peaceful Coexistence, which stand majestically around the tomb of the late Prime Minister Bandaranaike.

The colonnades have adopted the snowflake white marble veneer produced in China's Shandong Province, which means to bring Chinese cool snow to beautiful Sri Lanka, and solemn and beautiful BMICH. Noble gold chapiters on the colonnades support the wide overhanging eaves shining under the sun.

1

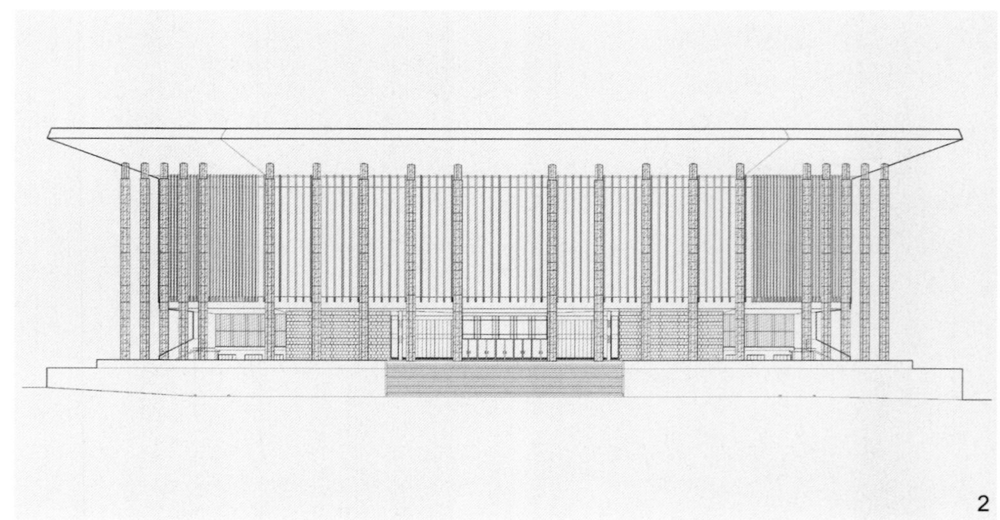

2

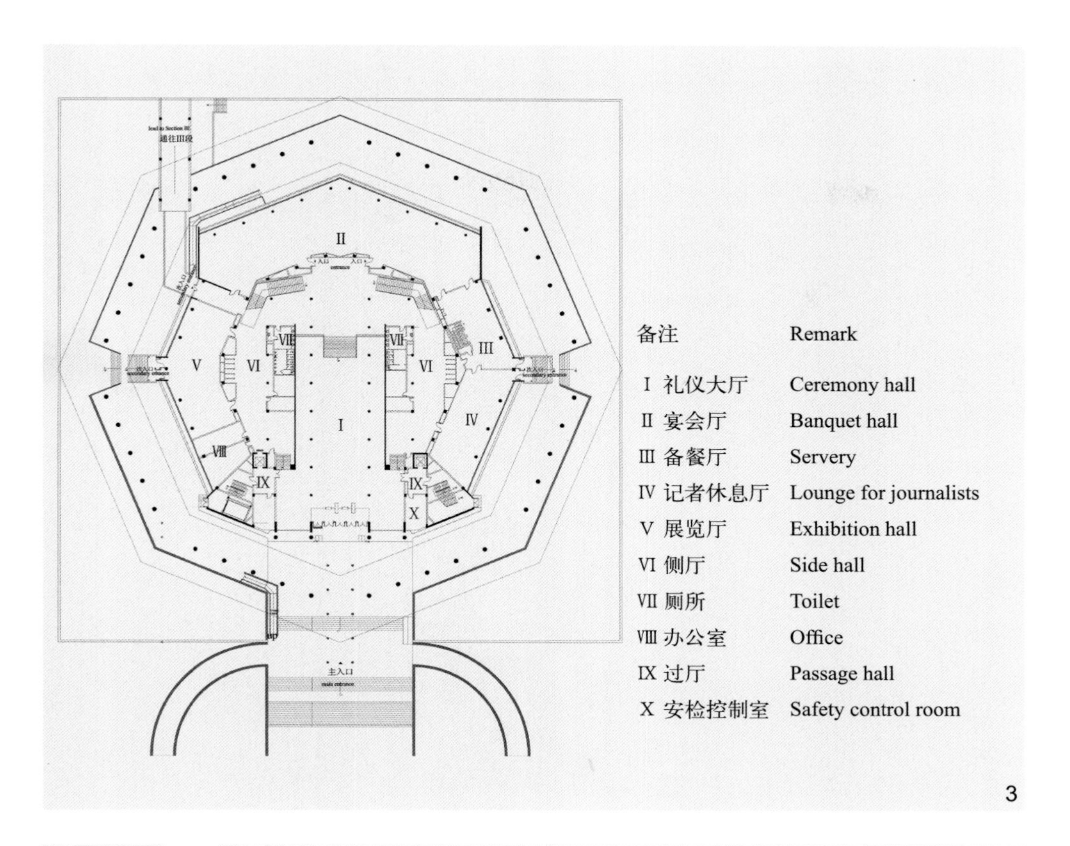

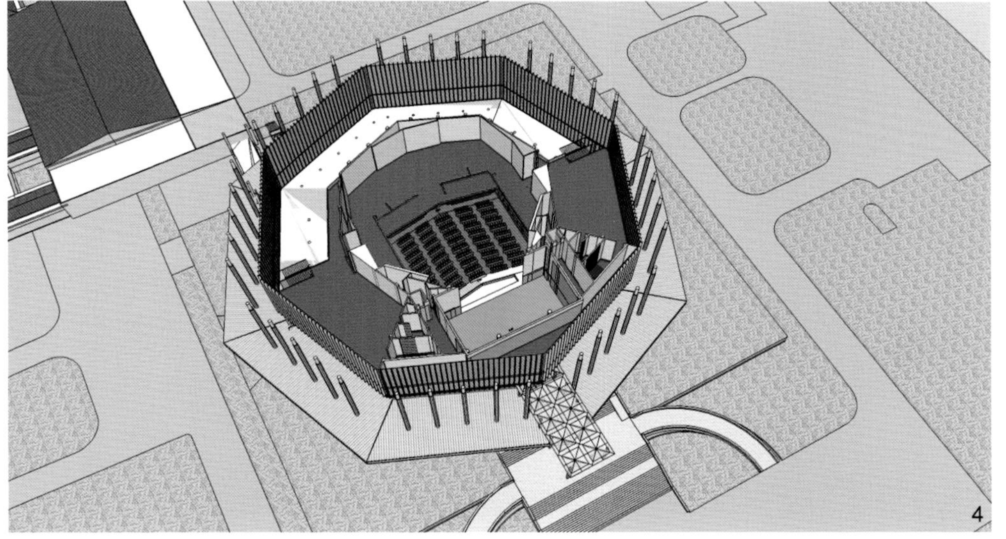

1. 晚霞柱子与白墙
   Columns with sunset patterns and white walls
2. 大会堂（一段）主入口立面
   Main entrance façade of Assembly Hall (Block I)
3. 一段一层平面图
   Ground Plan of Block I
4. 一段一层模型
   Model of ground floor of Block I

平顶屋面的设计更是很新颖，八角形屋面对角线长度为108米，巧妙地将108这个佛教吉祥数字融入建筑中，来突出斯里兰卡是一个有着悠久的佛教文化历史的国家。

建筑物内外均设置柱廊，充分强化其柱廊式的特点。大会堂内柱子装饰图案为晚霞，寓意美好的一天将要结束，更美好的明天将要来临，祝愿斯里兰卡人民永远幸福。

主体建筑地上建筑共3层，局部有夹层（5层）。

In the flat roof design, the octagonal roof's diagonal line is 108m long, which skillfully integrates the auspicious number of 108, an auspicious Buddhist number into BMICH, thus highlighting Sri Lanka's long history of Buddhist culture.

Colonnades, set both inside and outside the building fully highlight the building's features. The columns in the Assembly Hall are decorated with sunset patterns, implying that a beautiful day is coming to an end, a better tomorrow will come, and the wish of happiness forever for the Sri Lankan people.

The main building has a three-storey building, locally with interlayer (five stories).

1

一层由主入口进入，就到达礼仪大厅，另有侧厅、展览厅、宴会厅、银行、邮电、小卖部、记者休息厅及办公用房。

踏上礼仪大厅的红地毯，典雅、庄重、富丽堂皇的氛围令人感受至深；正前方汉白玉台阶上，正中央的黑色大理石基座上，安放着已故总理班达拉奈克先生的雕像；班达拉奈克夫人逝世后，夫妻雕像均安放在汉白玉台阶上，让人肃穆起敬。

伫立在大厅两侧的18根八角形晚霞柱子与斯里兰卡人民喜欢的白色大理石装饰的两侧墙壁，交相辉映，使得整个礼仪大厅显得既庄严，又明朗。

From the main entrance on the ground floor, one can access the etiquette hall, as well as the side hall, exhibition hall, banquet hall, bank, post service, shop, press lobby, and offices.

Upon setting foot on the red carpet in the banquet hall, one is surprised by its elegant, solemn and magnificent atmosphere; the bust of the late Prime Minister Bandaranaike is placed on a black marble base in the center of the white marble step straight ahead; the bust of Mrs. Bandaranaike is also placed here, evoking great respect.

Eighteen octagonal columns with the sunset patterns on both sides of the hall correspond to walls decorated with white marble, which is favored by the Sri Lankan people, making the etiquette hall solemn and bright.

1. 礼仪大厅全貌
An overall view of the etiquette hall

2. 描绘总理班夫人参加“班厦”建设的油画
Oil painting illustrating Prime Minister Sirimavo Bandaranaike participating in BMICH construction

3. 礼仪大厅标牌之一
Signage of the etiquette hall

4. 位于礼仪大厅正面的风景油画
Landscape oil painting in front of the etiquette hall

1. 休息廊及小卖部
Lounge area and shop
2. 平面呈蝶形的宴会厅
Butterfly-shaped design of the etiquette hall
3. 迎宾时的宴会厅
The banquet hall ready for guests

绕过雕像，映入眼帘的是由我国著名画家艾中信先生和两名助手，沿斯里兰卡海岸线考察后，在当地绘制的斯里兰卡锦绣河山的巨幅风景油画。

1972年6月，周恩来总理在百忙中还亲自审阅了大厅这幅大型油画的画稿。

艺术家笔下的这幅油画用笔遒劲、大气恢宏，且又细致入微。那冉冉升起的太阳，寓意对未来的无限美好满怀祈盼，可谓寓意深远……

绕过这幅巨幅风景油画便可进入宴会厅。宴会厅的平面呈蝶形，10根圆形柱子呈放射状布置，像是张开双臂在欢迎来自世界各地的宾客。

Behind the busts, a painting by famous Chinese painter Ai Zhongxin and two of his assistants greet visitors. Ai Zhongxin painted a large oil painting depicting the splendid landscape of Sri Lanka after making a trip along the country's coastline.

In June 1972, Premier Zhou Enlai personally reviewed the sketches for the oil painting.

The oil painting is powerful, grand but nuanced. The rising sun implies great expectations for the future, bearing a profound meaning...

After walking past the large landscape oil painting, visitors arrive at the etiquette hall. The hall has a butterfly-shaped design, with 10 circular columns arranged radially, looking like open arms welcoming guests from all over the world.

从油画两侧的楼梯可上到二层。二层为1500座的会议大厅，设有7种语言同声传译设备及供演出大型歌舞剧的现代化舞台设施。

休息厅位于会议大厅两侧。大厅使用的是高达13米，最具斯里兰卡建筑特色的通透落地玻璃百叶窗；高达5米的18盏吊灯，名为风车灯，这是建筑师戴念慈专为大厦设计的，意为把中国的凉爽清风带到美丽的斯里兰卡，带到庄严俊美的“班厦”。整个大厅给人一种凉爽休闲的气氛，时而还会有鸽子、鸟儿能从百叶窗飞进，落在美丽的风车灯上，更为休息大厅增添了别样情趣！

三层设有会议厅公众席及耳光室、储藏室等。

大会堂地下一层，设工作人员餐厅及机房。

There are staircases on both sides of the oil painting to access the first floor, which includes a 1,500-seat conference hall, which is equipped with seven-language simultaneous interpretation equipment and modern stage facilities for large-scale musicals and operas

On both sides of the conference hall, there are lounge halls equipped with 13m-high transparent floor-to-ceiling full-height glass shutters with distinctive characteristics of Sri Lankan architecture; 18 chandeliers, as high as 5 meters, are called wind lights, and they were designed by architect Dai Nianci for BMICH with the aim to bring cool breezes from China to beautiful Sri Lanka and solemn and elegant BMICH. The entire lounge hall has a cool and leisurely atmosphere, sometimes pigeons and birds fly into the hall from the glass shutters and sit on the beautiful wind lights, adding special flavor to the lounge hall!

Auditorium, fore stage side lighting and storage area are set on the third floor.

The basement of the assembly hall has a staff restaurant and plant room.

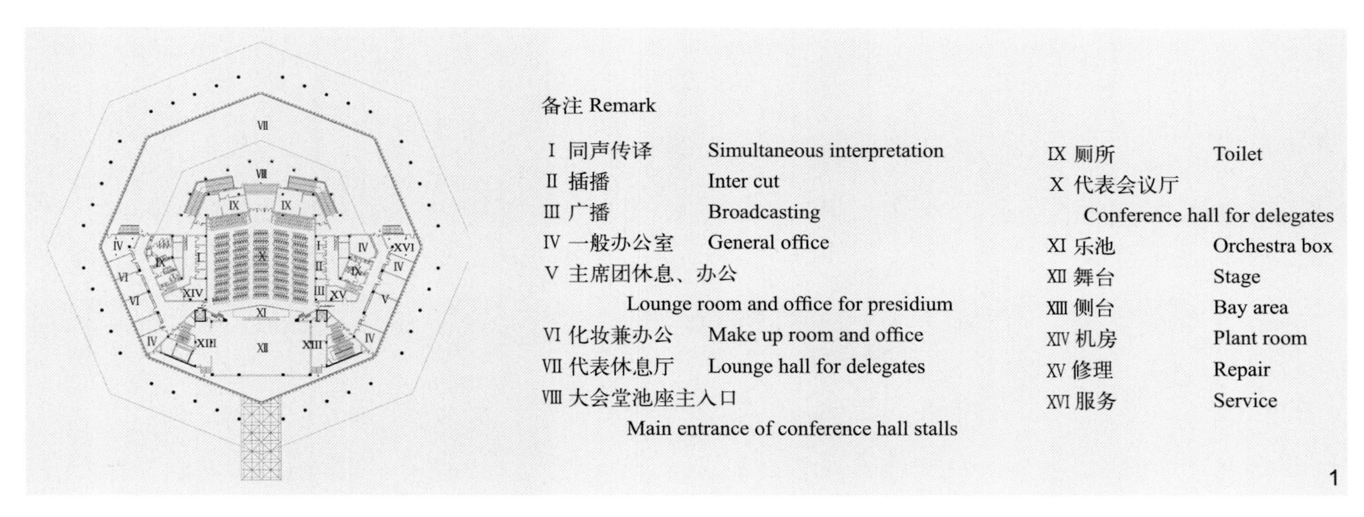

1. 一段二层平面图
Second Floor Plan of Block I

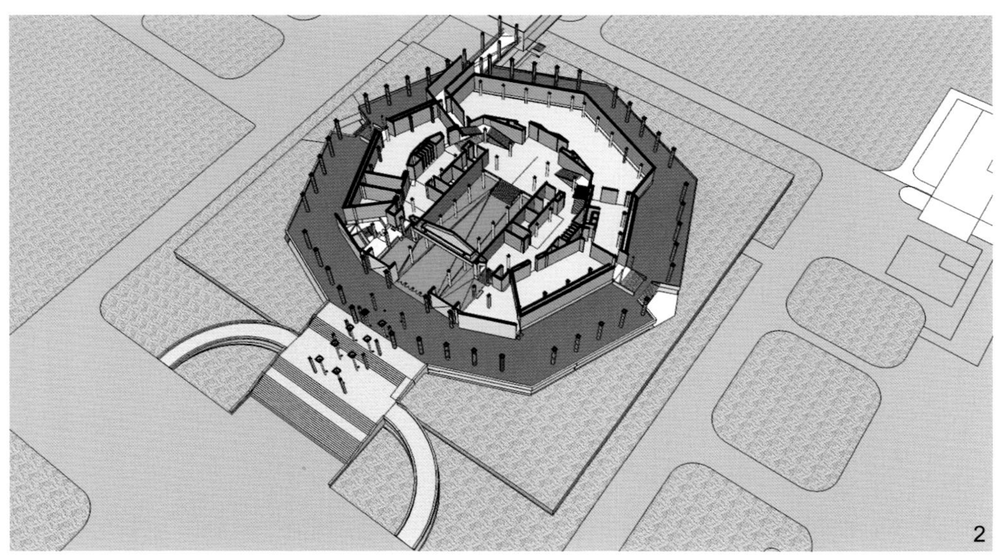

2. 一段二层模型
Model of Block I on the Second floor

1

2

3

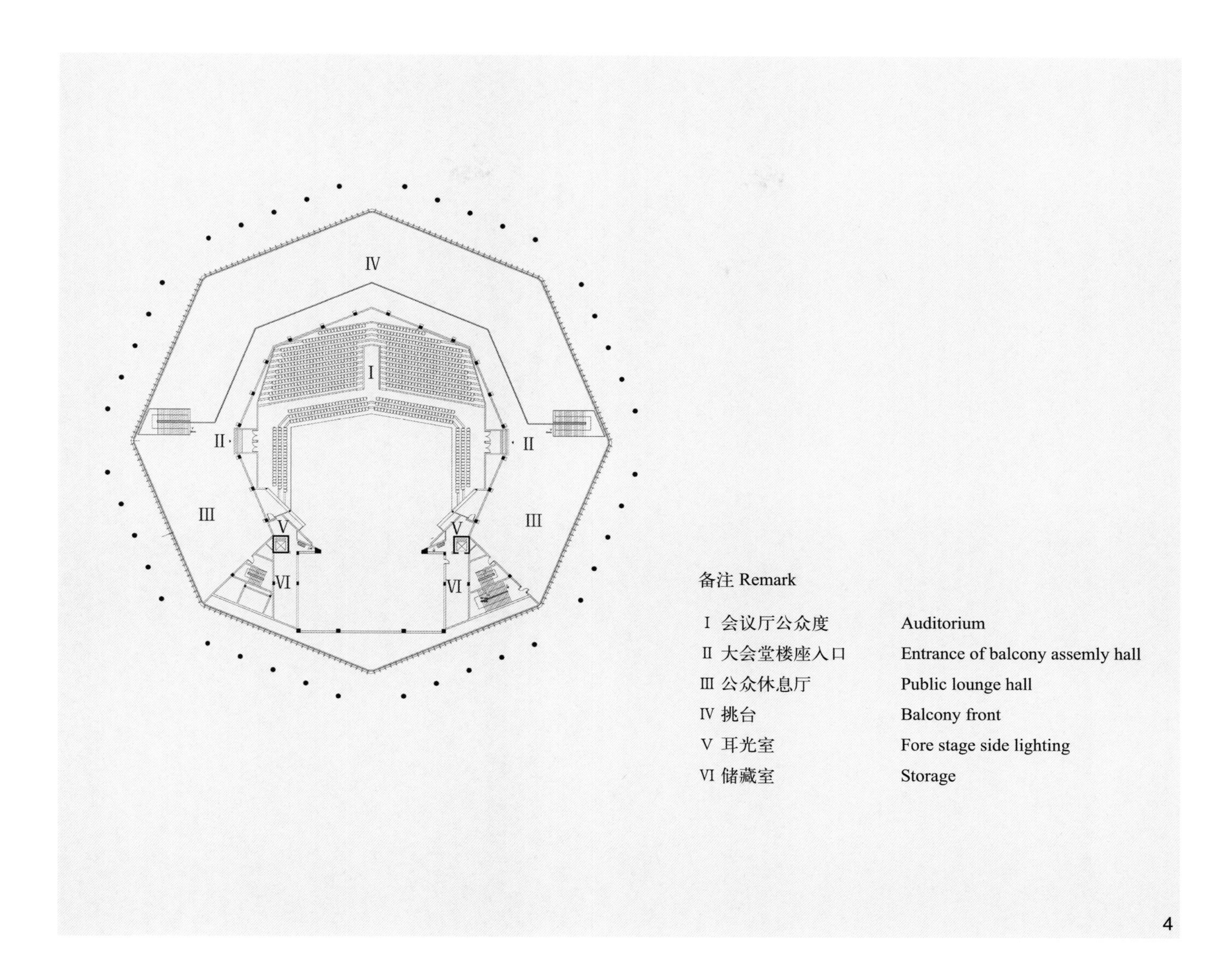

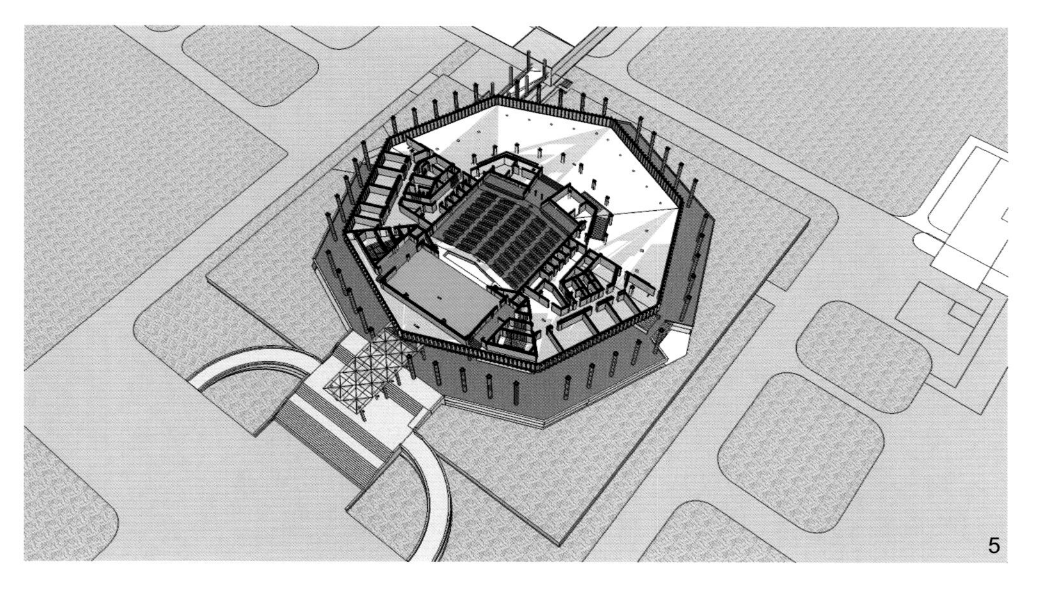

1. 休息大厅落地百叶窗、风车灯、折线楼梯全景
   A panoramic view of the floor-to-ceiling shutters, wind lights and folding stairs in the lounge hall
2. 休息大厅百叶窗
   Shutters in the lounge hall
3. 飞进来的鸽子落在美丽的风车灯上
   Pigeons sitting on a beautiful wind mill light
4. 一段三层平面图
   Third Floor Plan of Block I
5. 一段三层模型
   Model of Block I on the third floor

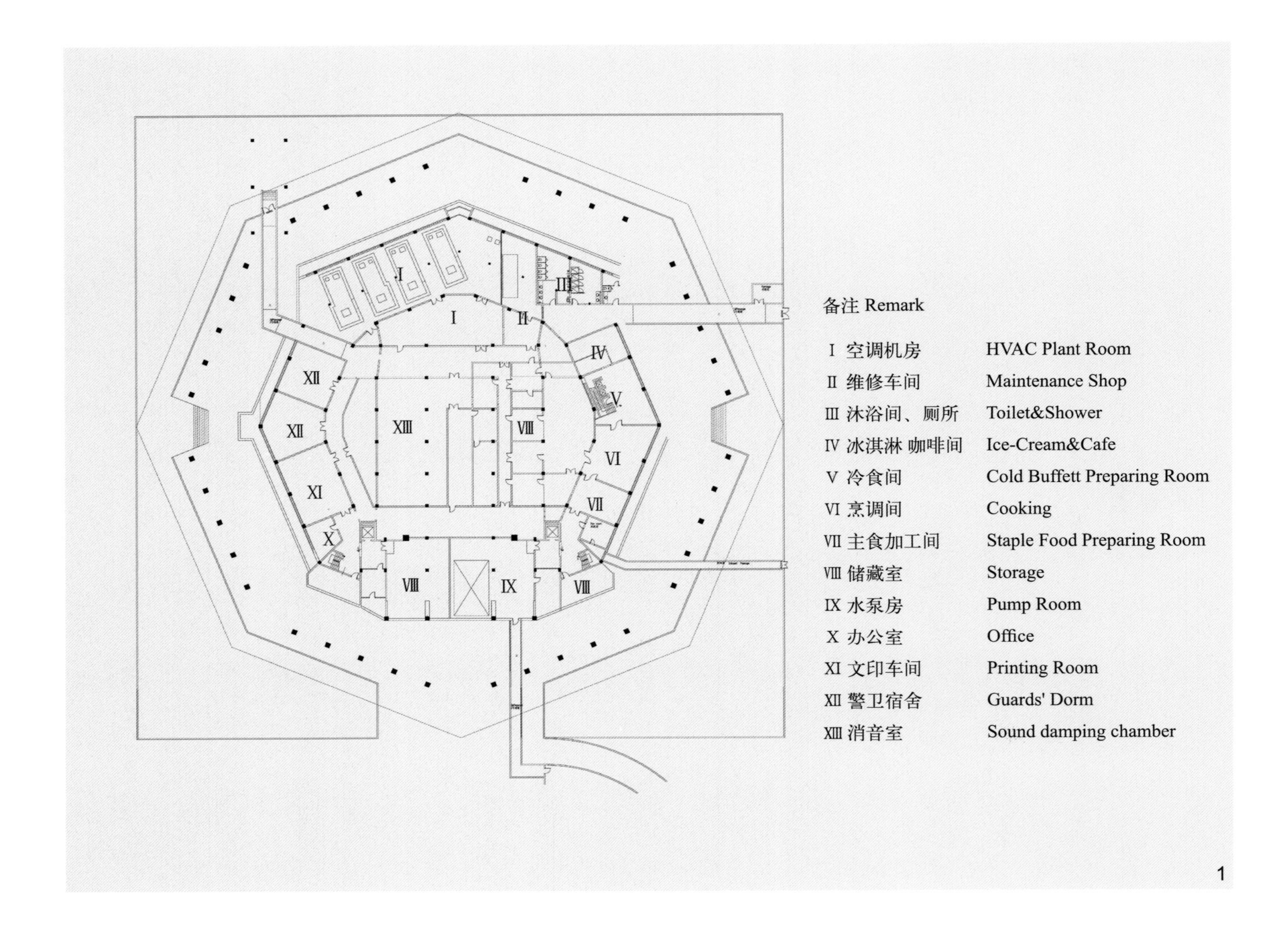

3. 附属建筑（二、三、四段办公楼群）

▪ 附属建筑总括

附属建筑与主体建筑通过主体建筑东南侧的一座天桥相连。附属建筑各单体建筑均由连廊、过道连接。

附属建筑主要用于中、小型会议和展览，大型国际会议服务等，各段均为矩形平面。二段是三层的可供90个代表团使用的办公楼；三段是两层的小会议室和讲演兼电影厅；四段则是三层的秘书处办公楼及工作人员餐厅。

3. Auxiliary Buildings (office buildings in Blocks II, III and IV)

▪ **Overview of auxiliary buildings**

On the southeastern side of the main building is an overpass connecting it to auxiliary buildings. The individual auxiliary buildings are connected to each other through connecting corridors and passageways.

The auxiliary buildings are mainly used for medium- and small-sized conferences and exhibitions, large-scale international conference services, and all the blocks are rectangular in design. Block II has three floors and is used by 90 delegations as an office building; a two-storey small committee rooms, and lecture and cinema hall are in Block III; the three-storey Block IV has the secretariat office building and staff restaurant.

1. 一段地下室平面图
Basement Floor Plan of Block I

2. 二、三、四段一层平面图
First Floor Plan of Block II,III,IV

3. 二、三、四段二层平面图
Second Floor Plan of Block II,III,IV

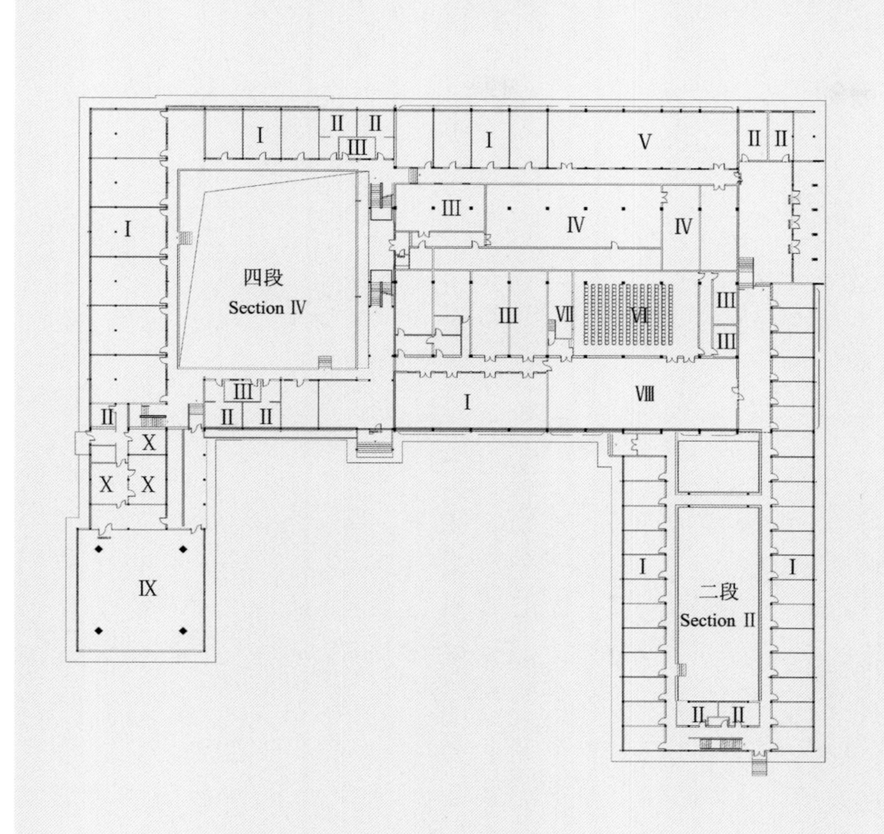

备注 Remark

Ⅰ	办公室	Office
Ⅱ	厕所	Toilet
Ⅲ	储藏室	Storage
Ⅳ	空调机房	HVAC Plant Room
Ⅴ	电力室	Power Room
Ⅵ	三段电影厅	Cinerna of Section Ⅲ
Ⅶ	放映厅	Screening Room
Ⅷ	休息厅	Lounge
Ⅸ	四段小餐厅	Restaurant of Section IV
Ⅹ	厨房	Kitchen

2

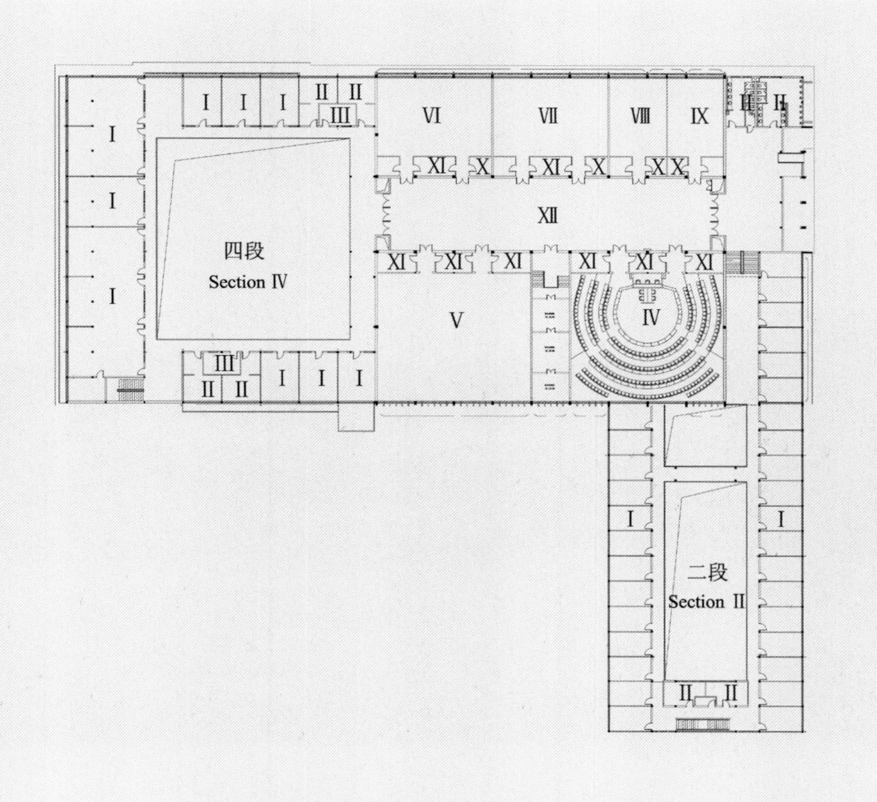

备注 Remark

Ⅰ	办公室	Office
Ⅱ	厕所	Toilet
Ⅲ	储藏室	Storage
Ⅳ	会议室A	Meeting Room A
Ⅴ	会议室B	Meeting Room B
Ⅵ	会议室C	Meeting Room C
Ⅶ	会议室D	Meeting Room D
Ⅷ	会议室E	Meeting Room E
Ⅸ	会议室F	Meeting Room F
Ⅹ	机房	Equipment Room
Ⅺ	备用室	Spare Room
Ⅻ	休息厅	Lounge

3

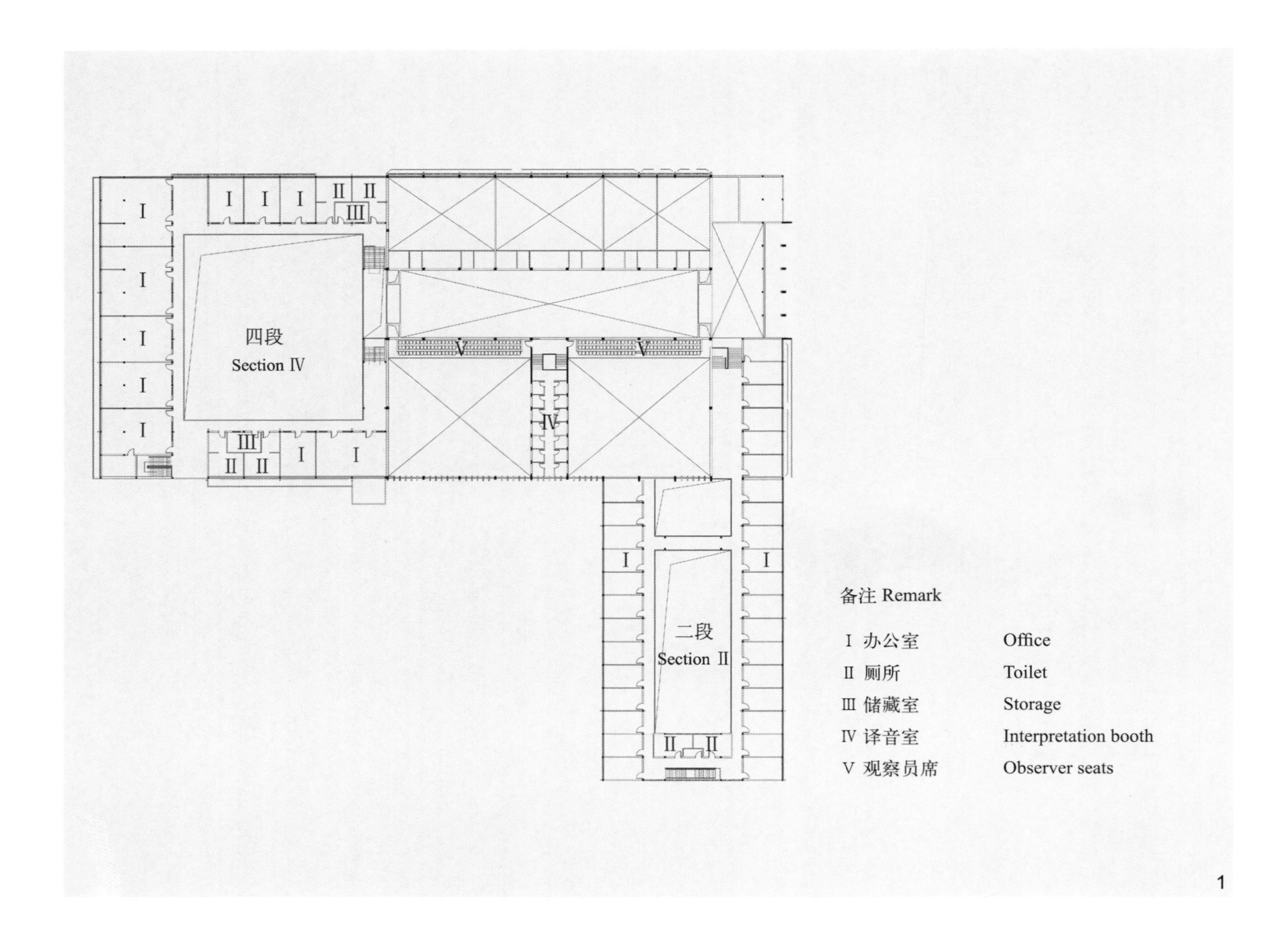

1. 二、三、四段三层平面图
Third Floor Plan of Block II,III,IV

2. 连接主体建筑与附属建筑的天桥
Overpass connecting the main building with auxiliary buildings

3. 代表团办公楼院内的天桥式连廊
Overpass corridor in the yard of the delegation office building

3

▪ 代表团办公楼（二段）

代表团办公楼为3层楼，均为办公室。

平面布置为内天井开敞走廊式，内廊式的设计使得各房间的行走距离最短。内廊宽度为1.8米，宽敞明亮，在其中行走有一种雅静、惬意的感觉。天井呈长方形，靠东侧的三层天桥式连廊将天井分为东西两部分，天井内绿草长青。首层宽度为0.4米的矮栏用浅蓝色马赛克贴面，可供人们休闲、乘凉。二、三层环形内廊均设金属栏杆，蓝色的栏杆、蓝色的金属门与白色的墙面形成鲜明对比。

天井的西侧设有一部楼梯，楼梯侧墙采用通透的预制水泥花格砖砌筑而成，被称为花格墙。且花格有向外倾斜度，以利排水。从一层至三层整面墙均为通透设计，除通风良好外，还有很好的装饰作用。

▪ **Delegation Offices (Block II)**

The delegation office building has three floors.

The delegation office building design features an internal courtyard and open corridor style, with the wide and bright middle corridor at 1.8m in width, which enables the shortest distance among all the rooms and gives people walking through the corridor a sense of serenity and comfortable feeling. The rectangular-shaped courtyard has evergreen grass, and the third-floor overpass corridor on the east side separates the courtyard into its eastern and western parts. The 0.4m-wide short railing on the ground floor that has light blue mosaics provides a place of leisure and relaxation. The circular middle corridors on the first and second floors are set with metal railings, and the blue railing, blue metal door and white walls create a sharp contrast with each other.

On the west side of the courtyard, there is a stairway, and on the side wall, there are transparent precast cement lattice bricks, creating a lattice wall. The lattice wall is designed at a certain outward angle to facilitate water drainage. The entire walls on the ground, first and second floors are transparent, which not only ensures good ventilation but also has very good decorative effect.

1

2

1. 代表团办公楼东、北侧（花格墙）
The east and north sides of the delegation office building (lattice wall)

2. 内天井开敞走廊
Open corridor in the courtyard

3. 蓝色的栏杆、蓝色的金属门
Blue railings and blue metal doors

4. 楼梯及花格墙
A staircase and lattice wall

屋顶降温采用架空层的做法，即在钢筋混凝土屋面上间隔砌筑砖垛，在砖柱上安装轻型钢檩条，其上安装瓦楞铁金属屋面，形成一个隔热、防水的效果。

沿办公室百叶窗上部还设有悬挑1.2米，下垂0.9米的加高遮阳板，再加上茂密的树木遮挡，室内完全可以达到舒适的办公环境要求。

▪ 委员会办公楼（三段）

委员会办公楼为2层楼，位于代表团及秘书处办公楼之间。一层设有一个讲演厅兼电影厅、办公室、储藏室、机房等。二层设有A-F六个会议室及办公室等。

To produce the roof cooling effect, an open-floor plan was adopted. Brick pillars were built at intervals on the reinforced concrete roof, and light steel purlins were installed on the brick column, with the corrugated iron roof on top, thus forming a thermal insulation and water-proofing effect.

Sun-shading visors were set above the office shutters, which overhang by 1.2m and hang down 0.9m, and together with dense trees, a comfortable office environment is realized.

▪ **Committee Rooms (Block III)**

The two-storey committee building is located between the buildings for the delegation offices and secretariat offices. The ground floor of the committee rooms is set with a lecture and cinema hall, office, storage, plant room, etc; the first floor is arranged with six committee rooms (labeled from A to F) and offices, etc.

1. 三段外景北立面
Northern façade of Block III

2. 三段外景
Exterior landscape of Block III

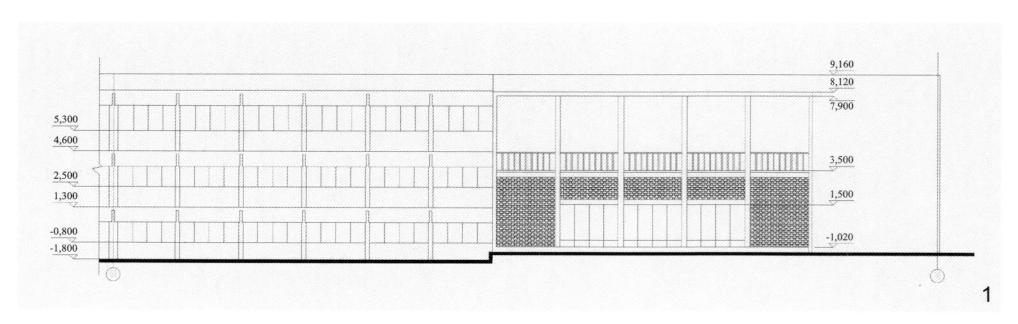

1. 委员会办公楼（三段）外遮阳板
External sun-shading visors of the committee offices (Block III)

2. 委员会办公楼（三段）主入口
Main entrance of the committee offices (Block III)

3. 讲演厅兼电影厅
Lecture and cinema hall

4. 讲演厅兼电影厅墙壁放映口及座椅
In-wall projection ports and seats in the lecture and cinema hall

1. 透过电影厅门厅的百叶窗，看到是2段里面。
Interior of Block II seen through the shutters in the cinema hall lobby

2. A（马蹄形）会议室
Committee Meeting Room A (horseshoe-shaped)

一层的讲演厅兼电影厅，可容纳208人。放映室内放置两台放映机及电影厅配电、灯光等控制柜，银幕后方设有库房。观众厅旁边，即为电影厅前厅——休息厅，面积共约300多平方米。

从前厅透过宽敞的玻璃百叶窗户看到的就是代表团办公楼的内天井，绿油油的草坪和来来往往的工作人员尽收眼底，是观众小憩的场所。

讲演厅兼电影厅内有独立的空调系统。委员会办公楼首层除了讲演厅兼电影厅外，东侧、南侧有窗户的房间用作办公室，居中部位的房间为各会议室中央空调主机系统等设备机房。

The lecture and cinema hall on the ground floor can accommodate 208 persons. The projection room is set with two projectors with power distribution and lighting control cabinets, as well as storage behind the screen. An antechamber, or lounge hall next to the auditorium has an area of 300m^2.

Through the spacious glass shutters from the antechamber, one can see the internal courtyard of the delegation rooms, the green lawn and working staff, as well as a place for audience members to rest.

The lecture and cinema hall has an independent air conditioning system. Apart from the lecture and cinema hall, rooms with windows on the east and south sides of the ground floor serve as offices, and rooms in the central areas have central air conditioning host systems for the committee rooms.

1

2

二层的A会议室设有438座，桌椅呈马蹄形布置，故也称马蹄形会议室。有七种语言同声传译设备，102个即席发言设备和扩声系统。每排桌后设有两排座椅，前排为可活动的贵宾椅，后排为固定的普通椅，利用圆弧内外弧长的差异半径大，1张活动椅子可匹配2张固定椅，较多地容纳了观众。会议室入口两旁为空调风管通道及电气控制间、储藏室。设计师充分利用净空，在这些设备用房的上层布置了可容纳50席的记者旁听席，并设有独立的记者出入口。地面均采用实木地板，利用台阶式地板下的空间，布置空调回风风管及风口，回风口布置在座椅下部，既隐蔽又满足空调送回风工艺要求。墙面均采用造型实木板条制作，具有吸声作用，木本色的桌椅与咖啡色墙面、白色吊顶、金色灯具等形成鲜明对比，尤显稳重、庄严。

B会议室设有438座，有7种语言同声传译设备，60个即席发言设备和扩声系统。地坪为本色实木小方块拼花地板，墙面同A会议室，没有安装固定桌椅，以利灵活布置，满足多种会议需要。吊顶为白色胶合板，安装有嵌入式方灯。同声传译分别设有7间同声传译室，房间位于A、B会议室之间，供A、B会议室使用，两室墙壁均设有窗口，便于观察会议情况。

Committee Room A on the first floor has 438 seats, and tables and chairs are arranged like a horseshoe, and it is thus named the horseshoe-shaped committee room, which is equipped with seven-language simultaneous interpretation facilities, 102 impromptu-speaking equipment and sound-reinforcement system. Two rows of seats are behind each row of tables, where the front row has adjustable VIP seats, and the back row has fixed ordinary chairs. Due to the large difference in the radius of the internal and external arch lengths, the combination of one adjustable seat and two fixed seats can accommodate more audience members. On both sides of the committee room entrance, there are the AC duct channel, electrical control room and storage room. The designer fully utilized clearance and arranged the 50 seats for journalists above the equipment room, with a separate entrance and exit for them. Solid wood floor was used, while the AC air return pipe and inlet is set below the terraced floor, and the air vent is concealed below the seats, which satisfies AC air supply and vent requirements. The use of solid wood on the walls allow for sound absorption, while the wood tables and chairs create a sharp contrast to the coffee-colored walls, white ceiling and gold lamps, thus demonstrating its dignified features.

Committee Room B has 438 seats, seven-language simultaneous interpretation facilities and 60 impromptu-speaking equipment and sound-reinforcement system. The floor adopts a natural colored solid wood small block floor, and the wall is the same as in Committee Room A, without fixed tables and chairs to allow for flexible arrangement to satisfy the demands of different conference layouts. The white plywood ceiling has an embedded square lamp. There are seven simultaneous interpretation rooms between committee rooms A and B. The walls of these two committee rooms are set with windows for observation.

1. B会议室
   Committee Meeting Room B
2. 座椅底下的回风口
   Air vent below the seat
3. 本色实木小方块拼花地板
   Natural colored solid wood on the small block floor

1

2

3

1. C、D会议室
Committee Meeting rooms C and D

2. 通廊两侧形成自然通风的花格金属门
Natural ventilation fence to the metal lattice doors on both sides of the corridor

C、D会议室形式相同，均设有50座，每个坐席设有即席发言设备。E、F会议室形式相同，均设有30 座。

六个会议室均设有中央空调系统。会议室之间设有一个宽敞的通廊，兼做休息厅，两端设有金属花格门隔断，通透的金属花格门设计，使得休息厅通风良好。

▪ 秘书处办公楼（四段）

秘书处办公楼为3层楼，3层均为办公室。

平面布置为内天井式，并带有环绕走廊。天井为正方形，内廊做法与代表团办公楼一样。天井中央有一棵修剪很精致的小树，四季常青，其环境优美，是一个天然氧吧。

Committee Meeting rooms C and D are the same style, both having 50 seats, and each seat is equipped with impromptu-speaking equipment. Committee Meeting rooms E and F are the same style, both having 30 seats.

All six committee Meeting rooms have central air conditioning systems. Spacious corridors between the committee Meeting rooms serve as lounge halls, with both sides partitioned by transparent metal lattice doors, which enable good ventilation in the lounge hall.

▪ **Secretariat Offices (Block IV)**

The three-storey secretariat building consists of offices.

The secretariat offices have an internal square courtyard, with a surrounding corridor. The internal corridor is the same as the delegation offices. In the center of the courtyard is a finely manicured small evergreen tree, and its beautiful environment is just like a natural oxygen bar.

秘书处办公楼（四段）内侧遮阳板

Internal sun-shade visors of the Secretariat Offices (Block IV)

秘书处办公楼正门前的草坪中央，竖有旗杆，其上飘扬着斯里兰卡国旗。每天早上均有升国旗仪式，当奏响斯里兰卡国歌的时候，这里的工作人员都会自觉地立正敬礼。

In the center of the lawn in front of the front gate of the secretariat offices, there is a flagpole waving the Sri Lankan national flag. Every morning, the national flag rising ceremony is held. When the Sri Lankan national anthem is played, the staff here stand up and salute.

秘书处办公楼（四段）入口
Entrance to the Secretariat Offices (Block IV)

1. 常青树和斯里兰卡国旗
The evergreen tree and the Sri Lankan national flag

2. 水磨石板矮栏及至今完好的马赛克地面
The terrazzo short railing and well preserved mosaic floor

3. 小朋友正在宽敞的矮栏上进行绘画比赛
Children's painting competition displayed on the spacious short railing

3

草坪与首层围廊用水磨石板矮栏分隔，围廊地坪采用灰色马赛克铺贴，矮栏水磨石板较宽，可供人闲坐。

在三层还设有总理班夫人专用的办公室和休息室。

在办公楼的东南角单独设有餐厅和厨房，供工作人员就餐。餐厅采用伞壳造型，与办公楼相对脱离，通过连廊联系。其独特的壳体结构除了满足功能需要，轻巧的造型也成为建筑群中的一景。

The terrazzo short railing separates the lawn and ground floor gallery, while the gallery floor is paved with grey mosaics, and the terrazzo short railing is wide enough for people to sit on it.

An office and lounge exclusively for Prime Minister Sirimavo Bandaranaike are on the second floor.

The staff restaurant and kitchen are located at the southeastern corner of the office building. The restaurant is shaped like an umbrella. It is separate from the office building, and connected with the office building by a corridor. Its unique umbrella-shaped design can meet functional demands, while its light shape also becomes a scene in the building complex.

1. 总理班夫人专用休息室
Lounge exclusively for Prime Minister Sirimavo Bandaranaike

2. 伞壳餐厅立面
Façade of the umbrella-shaped restaurant

3. 餐厅伞壳室内
Interior of the umbrella-shaped restaurant

4. 通向伞壳餐厅的连廊
Connecting corridor leading to the umbrella-shaped restaurant

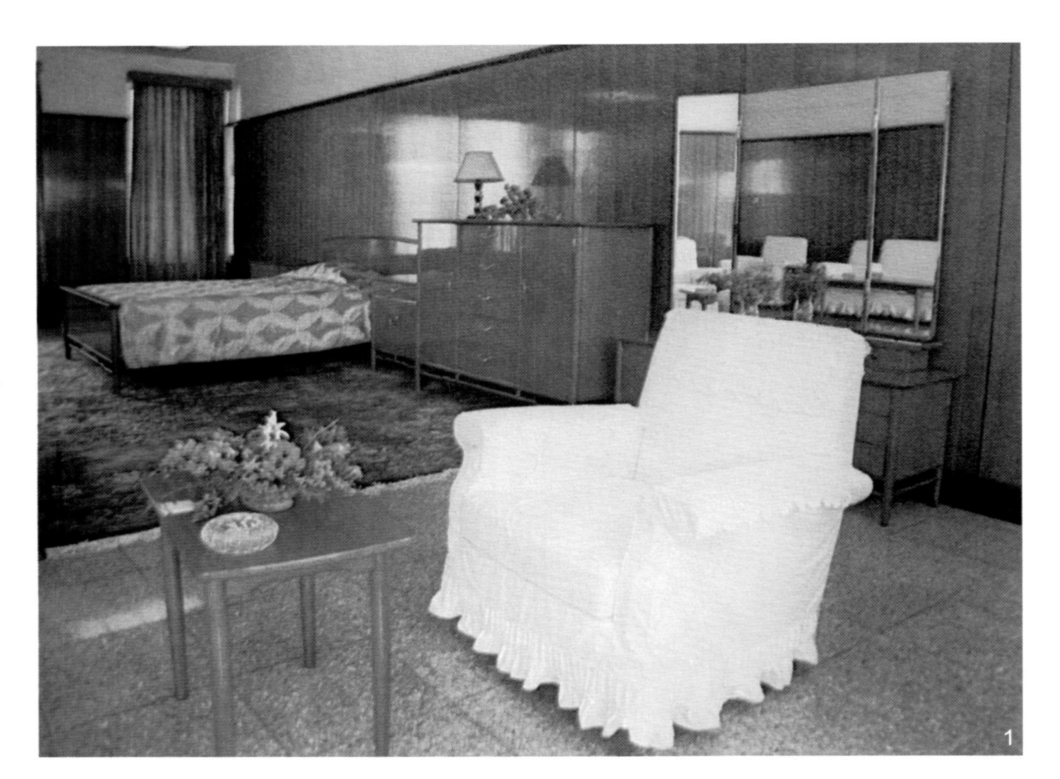

1

2

3

4

二 结构设计

1. 各段结构选型一览表

“班厦”大会堂、办公楼及其他配套辅助建筑等部分的结构概况如下表。

2. 屋面结构设计

第一，一段屋面是八角形的，对角线长度为108米，外檐挑出12米。外廊有40根24米 高的柱子，而内部柱网为八角形。在8米标高处，楼面通过由八角形柱网向外悬挑梁的办法形成内八角形楼面，而这个楼面才是“班厦”会议厅的地面。因此，屋面的施工必须是在8米标高钢筋混凝土楼板施工完成后进行，这是在设计时必须考虑的。

(II) Structural Design

1. List of structural type of each block

The structural overview of the Assembly Hall, Delegation Offices and other auxiliary buildings of BMICH are shown in the table below.

2. Roof structure design

Firstly, the roof of Block I was an octagonal shape, with the diagonal length was 108m with an overhang of 12m. The verandah had 40 pieces of 24m-high columns and the internal column grid was octagonal. At the 8m elevation, the floor formed an octagonal shape by the overhanging beams facing outward via the octagonal column grid, and this floor was exactly the same as the one in the assembly hall of BMICH. Therefore, the roof construction had to be done by constructing the reinforced concrete floor slab of 8m high, which had to be considered during the design stage.

位置	面积平方米	主要结构形式	主要建筑材料
大会堂（一段）	18000	框架结构，地下1层，地上3层，局部有夹层（5层），其中地下室局部设有混凝土剪力墙；基础为柱下独立基础及条形基础。屋架为桁架系统，有屋架支撑	基础及承重系统为现浇钢筋混凝土结构，混凝土标号200号（局部250号）；屋架为型钢桁架，杆件主要由双角钢和槽钢组成
代表团办公楼（二段）	2288	3层框架结构，柱下独立基础	现浇钢筋混凝土，标号200号
委员会办公楼（三段）	6160	2层框架结构，柱下独立基础。屋架局部为钢桁架，局部为混凝土屋顶附加金属板屋面	主体结构及部分屋架为现浇钢筋混凝土，标号200号；部分屋架为型钢桁架，杆件主要由双角钢和槽钢组成
秘书处办公楼及冷冻机房、污水泵房、配电机房等（四段）	5012+975	3层框架结构，柱下独立基础 单层框架结构，柱下独立基础	现浇钢筋混凝土，标号200号 现浇钢筋混凝土，标号200号

LOCATION	AREA (m^2)	MAIN STRUCTURAL TYPE	MAIN BUILDING MATERIALS
Assembly Hall (Block I)	18,000	Frame structure, 1 basement floor and three floors above ground, partial interlayer (five floors), partially equipped with concrete shear wall in basement; independent foundation below column and strip foundation; truss system supported by roof truss	Cast-in-situ reinforced concrete structure for the foundation and bearing system, with #200 concrete (partially #250); section steel truss for roof truss and elements consisting of double angle bar and channel steel
Delegation Offices (Block II)	2,288	Three-floor frame structure and independent foundation below columns	Cast-in-situ reinforced #200 concrete
Committee Rooms (Block III)	6,160	Two-floor frame structure and independent foundation below columns; partial steel truss for roof truss and partial concrete roof equipped with metal plate roof	Cast-in-situ reinforced concrete for main structure and partial roof truss, with #200 concrete; section steel truss for partial roof truss and elements consisting of double angle bar and channel steel
Secretariat Offices and refrigerating plant room, sewage pumping station and transformer house (Block IV)	5,012+975	Three-floor frame structure and independent foundation below columns; Single-floor frame structure and independent foundation below columns	Cast-in-situ reinforced #200 concrete

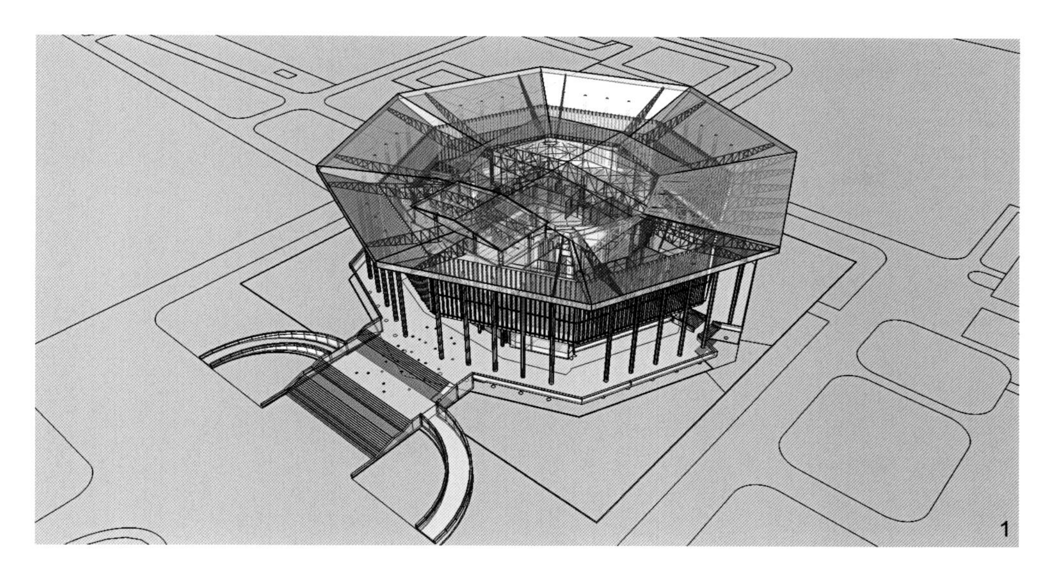

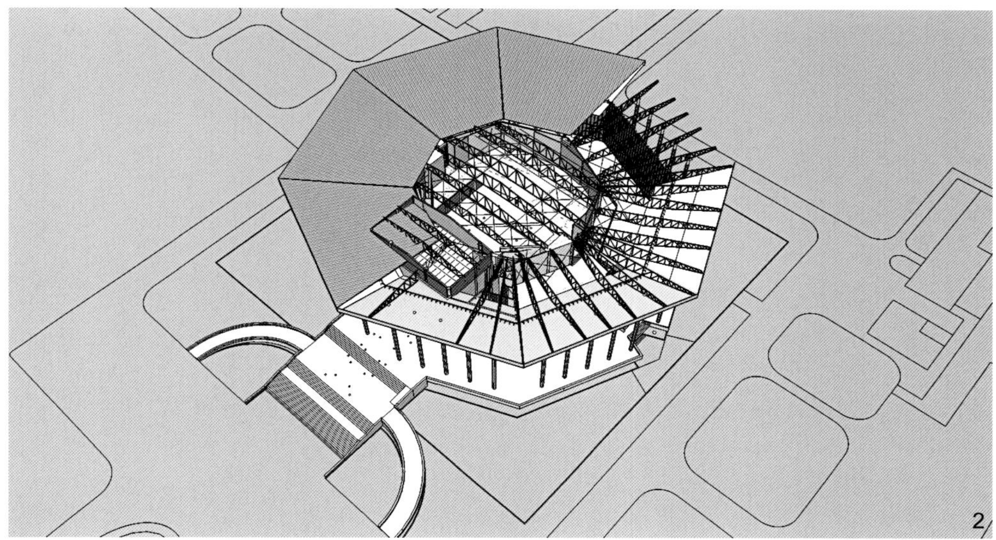

1. 一段屋面模型
Roof model of Block I

2. 一段屋面局部钢屋架模型
Partial steel roof truss model of Block I

3. 外挑钢屋架支撑简图
Schematic diagram of overhanging steel roof truss support

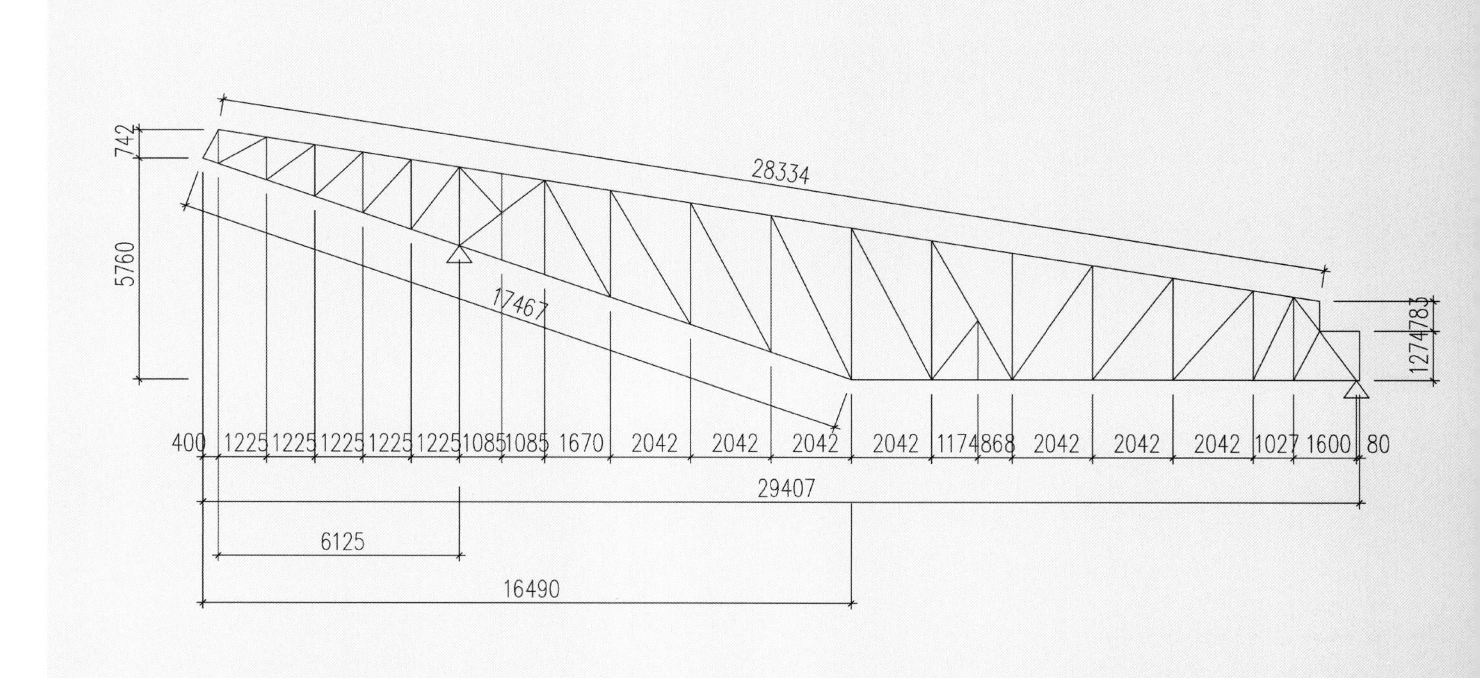

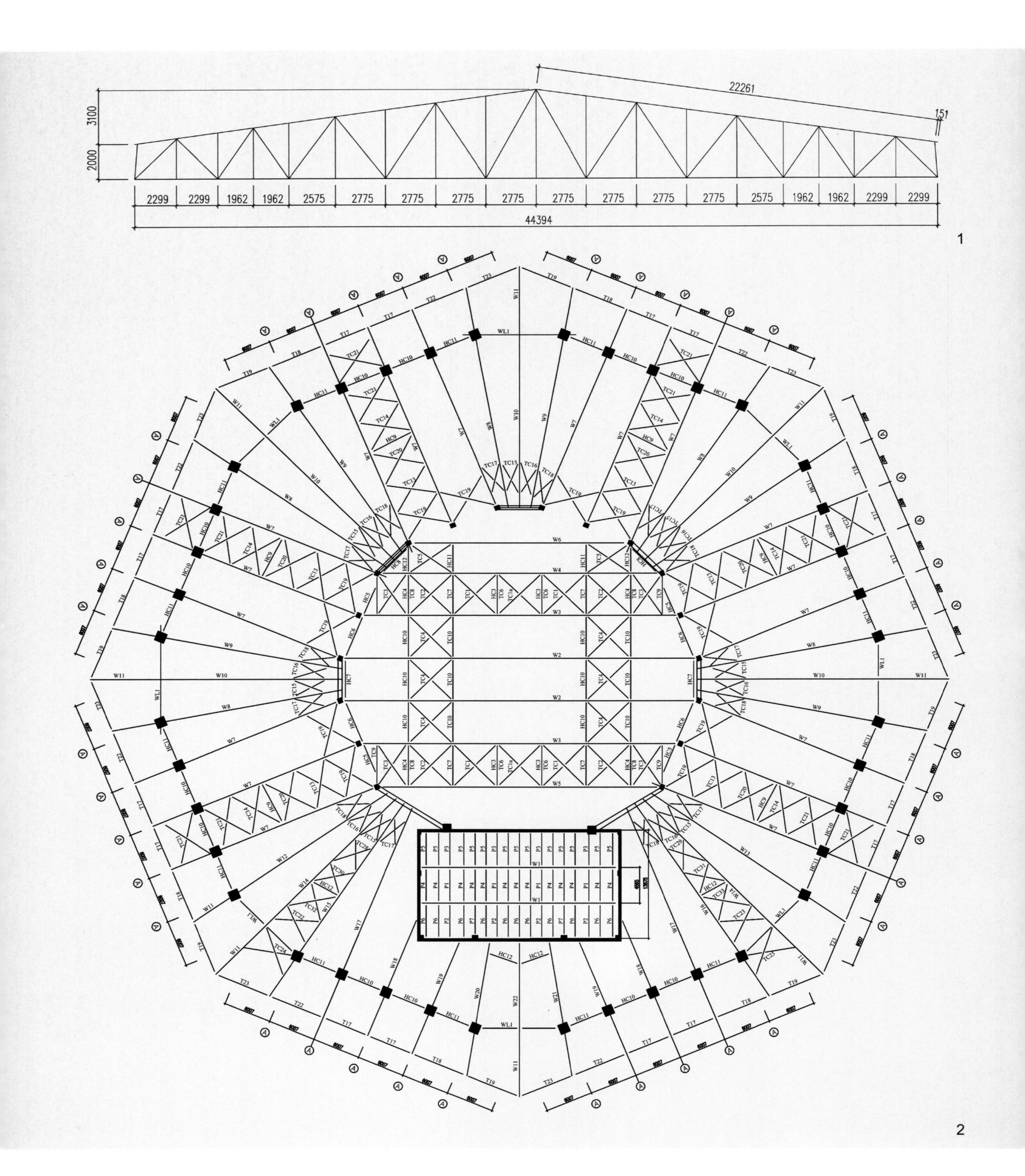

1. 观众厅屋顶钢屋架图
   Steel roof truss drawing of the auditorium
2. 一段钢屋架平面布置图
   Plan layout of steel roof truss of Block I
3. 一段屋顶钢屋架及马道
   Steel roof truss and catwalk of Block I
4. 角部两根外廊柱、托架与钢屋架的支撑关系
   Relationship between the two colonnades at the corners, bracket and steel roof truss support

第二，当时塔吊臂短，起重量低，为了结构轻便，结构专家采用了在建筑平面的八角形柱网上做平面梯形钢屋架的设计方案。

这个方案是首先利用八角形柱网最大跨度的平面梯形钢屋架的上弦，形成一道双向屋脊（简称：上弦屋脊）；由于平面钢屋架跨度逐步减小，使得矢高逐步递减，在矢高点两侧用两道檩条将矢高连接起来，就形成了两道与上弦屋脊成90度的矢高屋脊；这样，就自然形成一个中间突出，向四面坡下的十字屋脊。

之后，在四个面的每个面的中间，利用檩条，再形成四个檩条屋脊;且檩条屋脊的坡度，要与上弦屋脊和矢高屋脊的坡度一致；檩条屋脊的坡度是由檩托和檩托上的螺孔的位置来调节的。

最后，用平面梯形钢屋架制作八角形屋面的设计方案就完成了。

为了檩托和螺孔位置的精确，平面梯形钢屋架，是在国内加工好后，分段包装，运到“班厦”工地的。

Secondly, as the tower crane was short and has a low lifting capacity at that time, for a light structure, the structural experts adopted a design scheme to make a planar in a trapezoidal steel roof truss on the octagonal column grid of the building plane.

In this scheme, a two-way ridge is formed by using the upper planar of the largest-span plane trapezoidal steel roof truss of an octagonal column grid (short as upper chord ridge); since the rise gradually decreased, two purlins were used to connect the rise on both sides of the rise point, and two rising ridges form a 90-degree angle with the upper chord ridge; naturally, a cross-shaped ridge was formed, protruding in the middle and going downward along the slopes of the four facets.

Afterward, the purlins were used to form four purlin ridges in the middle of the four facets. The slopes of the purlin ridges adjusted according to the positions of purlin base and holes were screwed on it, making it consistent with the upper chord ridge and rising ridge.

Up to this point, the design scheme was completed.

To ensure the exact positions of the purlin base and screw holes, the plane trapezoidal steel roof truss was transported to the BMICH site segment by segment after it was produced in China.

钢屋架与柱子连接节点
Connection point of the steel roof truss and columns

1. “班厦”主入口8米标高挑檐
8m-level overhanging beam at the main entrance of BMICH

2. 大会堂主入口8米标高挑梁及遮阳板
8m-level overhanging beam and sunshade visor at the main entrance of the assembly hall

3. 8米标高楼板悬挑梁设计

在标高8米的结构平面上，有悬挑最长达7米、最短2米的钢筋混凝土梁，端部承载着4根高10米，间隔1.1米，重2.75吨的钢筋混凝土遮阳板。这样每根悬梁端部要承受约11吨的集中荷载，这无疑给结构设计带来了极大的困难。而且梁断面因建筑要求不能太大，但设计专家考虑：封檐梁刚度大，而这些悬挑梁与封檐梁共同形成一个空间体系，在结构的空间作用下，中间悬挑出7米的梁，变形肯定会小于实际计算数字。这样，就减小了梁的断面。

3. Overhanging beam design of 8m-level floor slab

On the 8m-level structural plane on which 2-7m of overhang reinforced concrete beams were set, the ends supported the four reinforced concrete sunshade visors that were 10m high, 2.75t and placed at 1.1m intervals. As a result each overhanging beam end supported about 11t of concentrated load, which made the structural design more difficult. Moreover, though the construction requirements asked for a limited beam section, the designers made the following considerations: the eaves beams were extremely stiff, and these overhanging beams formed a three-dimensional system with the eaves beams. Therefore, under the structural space, the formation of the 7m beams overhanging from the middle must be less than the calculated value. In this case, a reduced beam section was available.

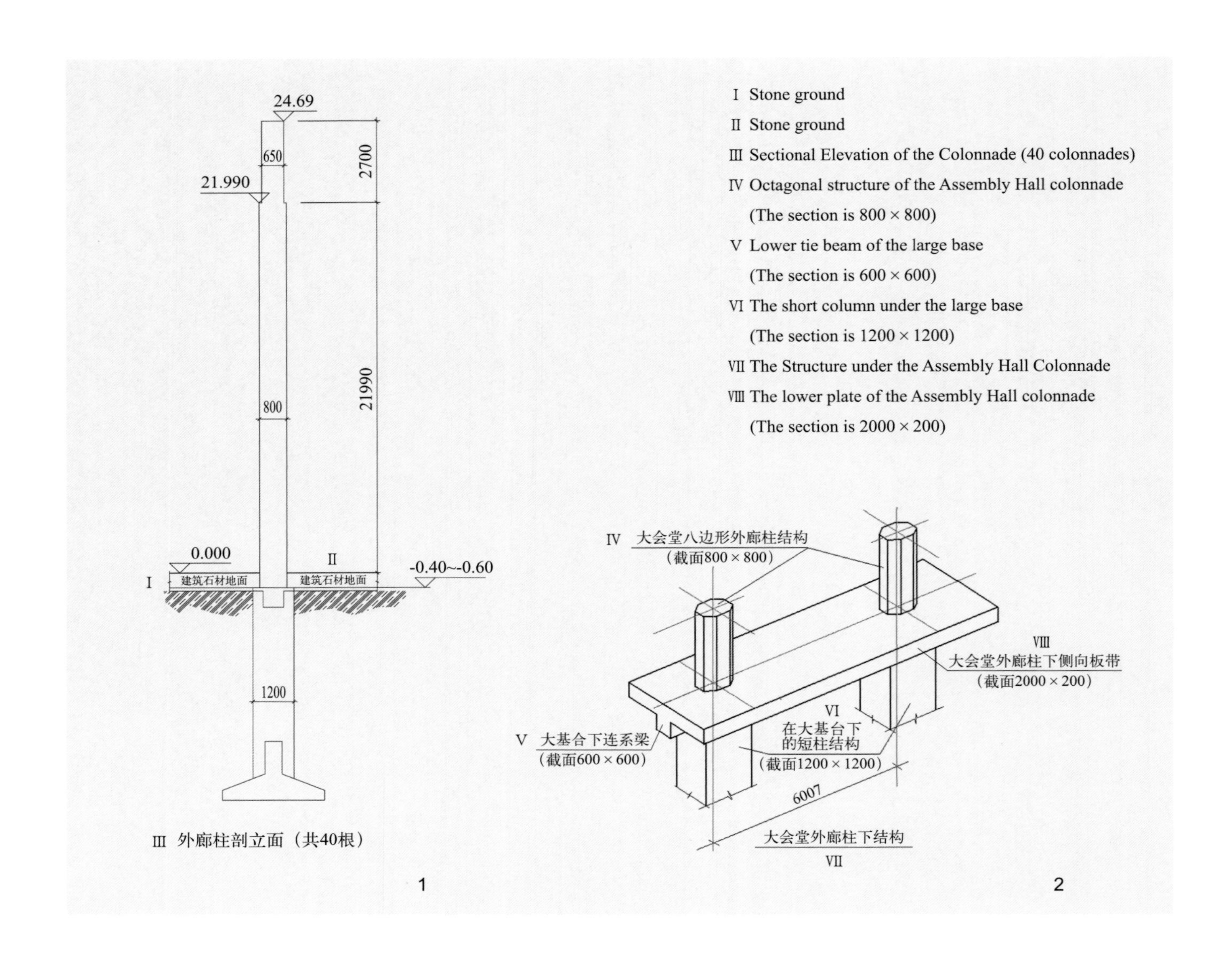

1. 外廊柱剖立面图
Elevation/section drawing of a colonnade

2. 外廊柱柱基构造示意图（侧向板带略）
Structural schematic drawing of a colonnade foundation (the lateral plate and strip is omitted)

4. 外廊柱子设计

建筑师在设计大厦外廊柱子时，改变了以往大型建筑柱子的粗壮形象，采用了纤细、挺拔、俊美、轻巧的形式。这就又给结构设计提出了特殊的要求，对柱子的结构设计既要保证建筑美观，又要确保安全。

“班厦”外廊柱子共有40根，呈八角形，与“班厦”主体外墙相距9米，环绕而成；柱子外露高度24米，但在“班厦”地下室部分有6.5米。因此，柱子总高度实际是30.5米。如按照常规设计，视觉上应是粗壮的形象。

4. Design of the BMICH Colonnades

When designing the BMICH colonnades, the architects tended to alter the previous stout image and presented a slender, robust, beautiful and light one instead. This resulted in special requirements for the structural design, which guaranteed a good appearance and safety for the column structural design.

There are 40 BMICH verandah colonnades, in an octagonal shape surrounding the main external walls of BMICH (9m from the walls). The exposed height of the columns reached 24m, but there was another 6.5m hidden in BMICH's basement, so the total height of the colonnades was actually 30.5m. Designed this way it has a stout look.

结构设计师根据“班厦”的整体结构，巧妙地利用了大基台台面下的混凝土垫层，在与柱子接触的地方加厚垫层，并配上钢筋，加连系梁，形成柱子的支点，减少柱子的计算长度，从而也就减小了柱子的断面。

这纤细、挺拔、俊美、轻巧的外廊柱，至今仍以独特的风采矗立在斯里兰卡的首都科伦坡那片热土上。

Based on the overall structure of BMICH, the structural designers skillfully utilized the concrete cushion below the large foundation. Where the colonnades were attached, they applied thick cushioning, reinforcement, and connecting beams to form colonnade pivots, thus reducing the calculated length and the colonnade sections.

The slender, robust, beautiful and light colonnades, with their unique style, still stand on the land of Colombo, the capital of Sri Lanka.

5. 基础结构设计

基础采用钢筋混凝土满堂反梁基础，底板厚度为20厘米，梁高为100厘米，按弹性基础梁设计，地下室外墙为钢筋混凝土挡土墙。基础反梁上面安装钢筋混凝土预制板，板上再浇注混凝土成地下室地面层。外圈标高，做成钢筋混凝土条形基础。

5. Structural design of foundation

The foundation adopted a reinforced concrete reversed-beam mat foundation, with a floor thickness of 20cm and beam height of 100cm, and it was designed as an elastic foundation beam. The basement exterior wall was made of a reinforced concrete retaining wall; reinforced concrete precast slabs were installed on the foundation reversed beam; the concrete was poured onto the slab to form the basement's floor, elevating the outer ring to make a reinforced concrete strip foundation.

满堂基础图

Mat foundation drawing

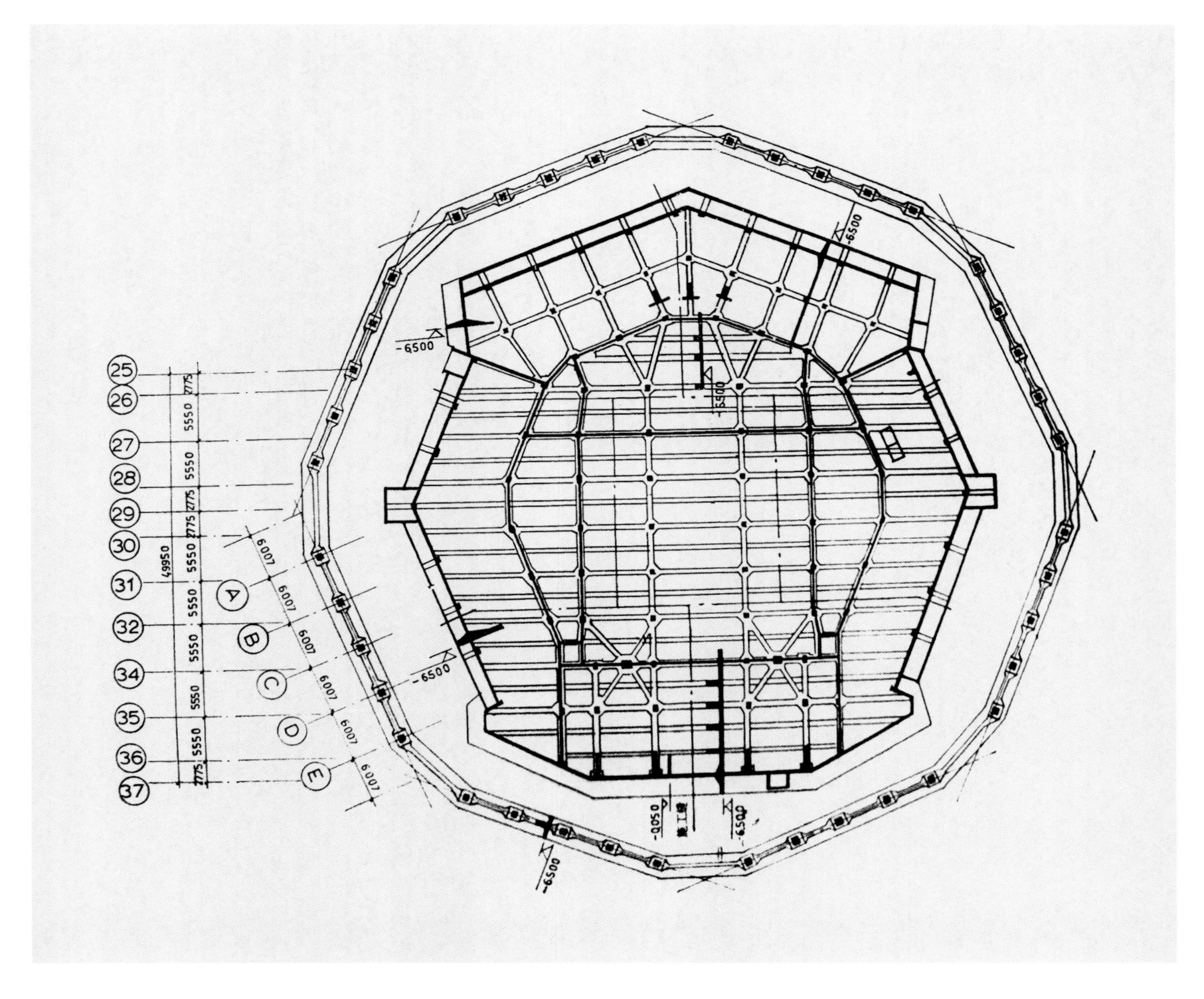

1. 伞壳餐厅立面设计图
Elevation drawing of the umbrella-shaped restaurant

2. 地基载荷试验的检验示意图
Schematic diagram of the foundation load test

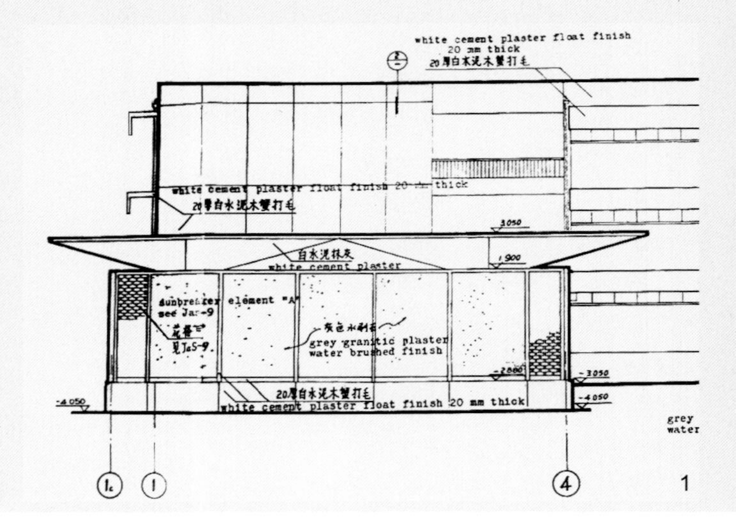

1

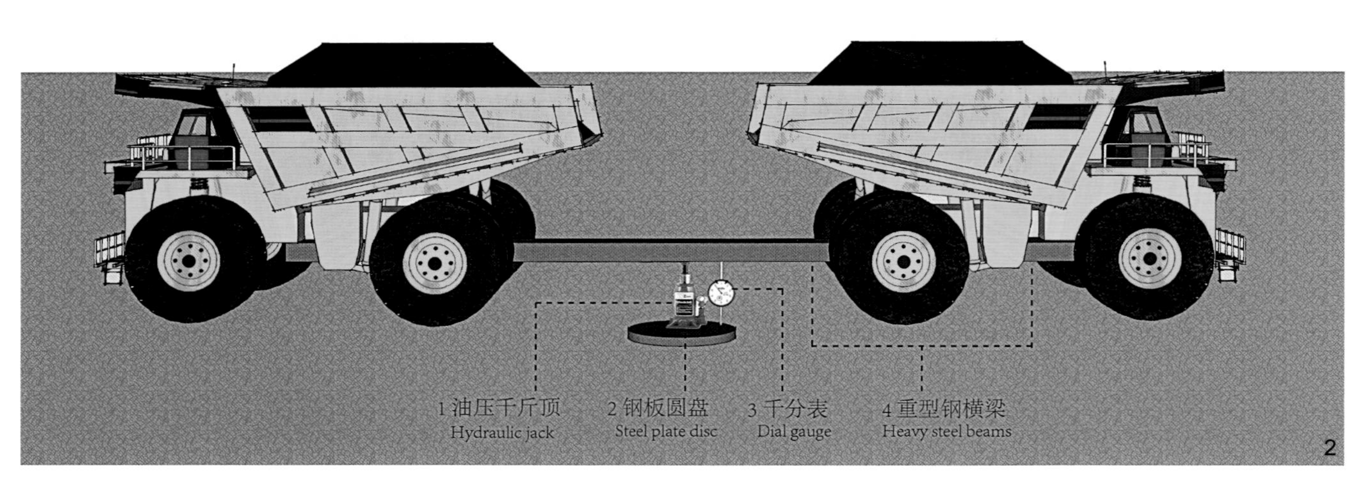

2

6. 伞壳屋面设计

伞壳餐厅的屋面，由4个12米×12米四边形伞壳组成。四边形的里边转角处都是一个梁，然后传到柱子上，从壳板面上看是一个抛物线，但模板可以用直线来做，比较简单，造型又非常美观。柱子底下做的是钢筋混凝土独立基础。

7. 地基载荷试验的检验

结构设计一开始，斯方寄到中国的地质勘探报告，数据不确切，工程师就根据考察时，挖探坑试验和以往的经验，确定地基承载力为180千帕（当时称地耐力为18吨/平方米），但心里一直不踏实,等到大厦开工后，基础土方挖完，就在基坑的基底面上，做了5个野外地基载荷试验，对地基承载力做了验证，验证证明，180千帕数据是稳妥的。做野外地基载荷试验时，由于当时没有锚桩，就用两辆卡车代替了。

6. Design of umbrella-shaped roof

The umbrella-shaped roof of the restaurant comprised of four 12m×12m quadrilateral umbrella shells. Since the internal nook was from a beam toward the column, the shell plate surface appeared to be a parabolic curve, but in the template a straight line could be used, which made it simple and beautiful. The reinforced concrete independent foundation was applied under the columns.

7. Inspection of the foundation test

Since the data in the geological survey report (sent by the Sri Lankan side) was not accurate from the beginning of the structural design, there was lots of uncertainty, as the engineers determined the foundation bear capacity as 180 Kpa (which was called $18T/m^3$ earth bearing capacity at that moment) based on trial pit test and previous experience. When construction started on BMICH and the foundation earthwork was finished, five field foundation load tests (as shown in the diagram) were conducted on foundation surface of foundation pit to verify the former figure. And it was proven that 180 Kpa was correct. It is noted that two trucks were used during the field foundation load tests due to a lack of anchor piles.

三　设备设计

如果说供电是“班厦”的动脉系统，那么供水就是“班厦”的呼吸系统，二者缺一不可，而且都在任劳任怨、“默默有闻”地辛勤工作着。“有闻”是它们正常的工作状态，变压器、发电机会有声音，冷水机组、冷却塔、水泵等也会有声音。为了让它们发出的噪声不影响大厦的工作环境，设计师将它们布置在距大厦适当的位置。并为它们设计自己的房间，就是变电室和制冷机房。

为了确保大厦供电的正常运转，配置了双回路电源供电，还有一台备用发电机。

给水排水系统则担负着大厦的生活、工作、卫生、消防等功能。

1. 电力设计

▪ 变电室设计

这座平房，就是“班厦”的变电室，三扇铁门所属的三个房间里各有一台变压器，两台1000kVA 和一台650kVA 油浸变压器，其中右侧的在建设初期设置的是一台30kW的柴油发电机，后来改为1000kVA 的柴油发电机。

(III) Mechanical Design

If the power supply is the arterial system of BMICH, then the water supply is the respiratory system of BMICH. Both are vital and willingly bear the burden of hard work with sound. This refers to the noise from the transformer and generator, the water-cooling unit, cooling tower and water pumps during the normal working mode. To prevent the noise from affecting the working environment in BMICH, the designers arranged for these machines to be placed away from BMICH and designed separate rooms for them, namely the transformer room and refrigerating plant room.

To ensure the smooth operation of power supply in BMICH, a two-circuit power and a standby generator were supplied.

The plumbing system is responsible for the life, work, sanitation and fire protection in BMICH.

1. Electrical Design

▪ **Design of transformer room**

There is a one-storey house for the transformer room of BMICH. There are three transformers in three rooms behind three iron gates, including two 1,000kVA and one 650kVA oil immersed transformers, and the one on the right was originally a 30kW diesel generator during the initial construction period and was later changed to the present 1,000kVA one.

变电室
Transformer room

1. 变电室高（右侧）低（左侧）压配电柜
   High voltage (right) and low voltage (left) distribution cabinets in the transformer room
2. 舞台灯光配电柜
   Stage lighting distribution cabinet
3. 舞台调光变压器及调节轮
   Stage dimming transformer and adjusting wheels
4. 手动调光机正面
   Front view of manual dimmer
5. “班厦”的工人师傅最后一次操作机器工作
   The last operation by a BMICH worker

▪ 舞台调光室设计

图中左边两台是电源柜，右边一台是自动调节柜，中间两台是灯光回路接线插孔连接柜。

舞台灯光配电柜设有90个回路插孔可以与观众厅和舞台灯光切换，还设有两个电源柜、一个自动调节柜。

舞台调光系统位于舞台下的夹层内，由三部分组成：舞台灯光配电盘、舞台调光变压器、舞台灯光操作。调光控制系统为手动控制。

▪ **Design of stage dimmer room**

On the left are two power cabinets, on the right is an automatically adjustable cabinet, and in the middle are two lighting loop patch connector cabinets.

The stage lighting distribution cabinet was equipped with 90 loop patch connectors to enable a lighting switch with the auditorium and stage, and with two power cabinets and one automatically adjustable cabinet.

The stage dimming system, located in the interlayer below the stage consists of three parts: the stage lighting distribution board, the stage dimming transformer and the stage lighting operation and adopted manual control.

▪ 扩声系统设计

大会堂观众厅扩声控制系统位于池座一侧的控制室，设有控制台、同声传译控制系统、监听设施等，并设有面向观众厅的瞭望窗口。扬声器主要置于舞台口上方及两侧。

▪ 电影放映室设计

电影厅放映室有2台座式35毫米放映机和配电柜、控制柜等，室内设有中央空调系统，观众厅的空调送风系统从放映室顶部的喷口吹出，回风通过座椅下的回风口，沿地坪下回风地道，经过放映室地坪下风道至空调机房。

▪ **Design of public address system**

The public address system in the auditorium of the assembly hall is located in a control room at one side of the seats of the orchestra and equipped with a control panel, simultaneous interpretation control system, monitoring facilities and observation windows facing the auditorium. The loudspeakers are largely set above the stage arch on both sides.

▪ **Design of projector room**

The projector room has two seat-type 35mm projectors and a distribution cabinet, control cabinet and central air conditioning system. The AC air supply system in the auditorium has the air conditioning distributed from the top vent of the projector room, and the air goes along the return air vent below the room through the air vent beneath the seats and towards the air conditioning plant through the air duct below the projector room .

1. 声控室
   The sound control room
2. 位于舞台上口的扬声器
   The loudspeaker at the back of the stage
3. 座式35毫米电影放映机
   Fixed-type 35mm film projector
4. 电影胶片电动倒带机
   Electric filmstrip rewinder

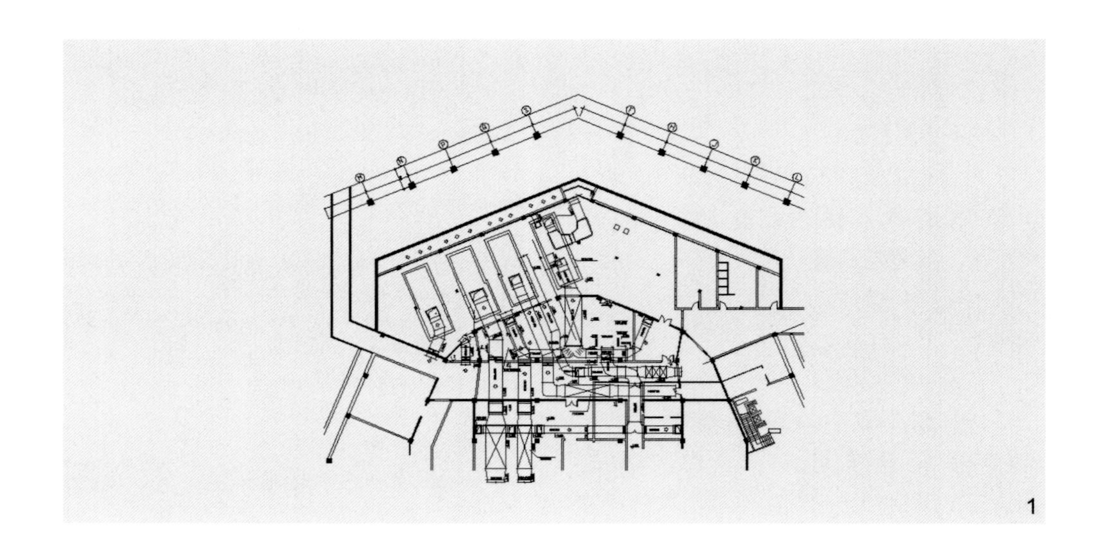

1. 一段地下室空调风管平面图
Plan of AC air duct in the basement in Block I

2. 观众场间休息大厅的落地百叶窗
Full length louvers at the audience lounge

3. 遮挡太阳辐射的遮阳板
The sunshade visor deflects the sun's rays

4. 大会堂蝶形瓦楞铁屋面及筒式通风孔
The butterfly-shaped corrugated iron roof and cylinder vent at the assembly hall roof

5. 在屋架上面直径800毫米筒式通风孔
The cylindrical vent on the roof truss with a diameter of 800mm

6. 耐久有效、通风简单的设备房
Durable and effective equipment room with simple ventilation

2. 通风、空调系统设计

在热带地区，公共场所的通风、空调显得尤为重要，“班厦”的通风、空调设计充分利用建筑构造实现尽量利用自然通风来满足使用要求，这种思路在“班厦”都得到充分的运用。

▪ 大会堂区域的设计

大会堂按照建筑功能序列可分为四个分区：礼仪大厅、宴会大厅、休息大厅、观众厅。观众厅和宴会厅设计采用集中空调，礼仪大厅和休息大厅则巧妙地利用建筑的自然通风，创造了良好的活动环境。

礼仪大厅入口两侧是落地百叶窗，末端的两个折线楼梯通道上面是大空间的休息大厅，楼梯起到烟囱效应，使气流自然穿过礼仪大厅形成穿堂风，起到降温的作用。

休息大厅有1500个座位供大会堂观众进行场间休息。有6个面是外窗、层高13米的大空间，使用了遮阳板和落地百叶窗，在屋架上还设有4个直径800毫米 的筒式通风孔，气流会通过休息厅的吊顶板缝隙进入屋架上空，观众厅周围的墙面到屋架上部6个面都留有孔道。

遮阳板起到反射太阳辐射热的作用，百叶窗和屋面通风孔形成大的自然对流，让偌大的休息大厅微风不断、凉爽宜人。

2. Ventilation and AC system design

Due to the tropical climate, the ventilation and air conditioning of public places are especially important, and for BMICH, this was achieved by maximizing the use of natural ventilation to satisfy the operating requirements through its construction. This practice was prevalent in BMICH.

▪ **Zoning design of the assembly hall**

The assembly hall can be divided into four zones according to building function sequences, including etiquette hall, banquet hall, lounge hall and auditorium hall. In them, central air conditioning system is applied to the auditorium and banquet halls, and the other two halls skillfully use the building's natural ventilation, thus creating a favorable active environment

Floor louvers are located on both sides of etiquette hall; a large lounge is located above the two staircases, with the stairs creating a chimney effect, by which the airflow passes through the hall to form a kind of cross ventilation and lowers the overall temperature.

The lounge hall accommodates 1,500 persons and has six facets equipped with external windows, a vast space with a height of 13m, sunshade visors, floor louver and four cylinder vents with a diameter of 800mm on the roof truss. The air flows to the overhead space of the roof truss by the ceiling board joining the lounge. Ducts were reserved from the wall space in auditorium to the six facets above the roof truss.

With the combined action of the sunshade visor (cutting solar radiant heat) and louver and roof vent (forming a natural convection), the large lounge hall was cool and pleasant, with a constant breeze blowing.

2

3

4

5

6

▪ 制冷机房设计

制冷机房的设计既独到又好用，用茂密的树木将其隐藏，既绿化了环境，又起到遮阳的作用，而且，还能掩盖制冷机发出的噪声。

“班厦”的制冷设备是我国第一台自己生产的制冷设备。

▪ 冷却塔机房设计

▪ 空调机房设计

空调室置于大会堂地下室和委员会办公楼首层，位置隐蔽且距离所供区域较近，风管沿吊顶、夹层、夹墙或地坪下布置，在利用建筑死角的同时满足通风需要。

▪ **Design of refrigerating plant room**

The refrigerating plant room had a unique and effective design, which made it invisible by lush trees, aiming to green the environment, keep away the sunshine and hide the noise from the machine.

The refrigerating equipment in BMICH was the first one manufactured by China.

▪ **Design of cooling tower plant room**

▪ **Design of the AC plant room**

The AC room, situated in the assembly hall's basement and on the ground floor of the committee offices, has a hidden location and is close to the service area. The air ducts are laid along ceiling, the interlayer, interwall or below the terrace, utilizing the building's dead space and meeting ventilation needs.

制冷机房内我国生产的第一台8FS-12.5型制冷机

The first 8FS-12.5 refrigeration machine in the refrigerating plant room manufactured by China

1. 这个隐蔽在绿色植物中的绿色房子就是冷却塔机房.
   Hidden among the green plants is the green house where the cooling tower plant room is located.

2. 空调间离心通风机
   Centrifugal fan in AC plant room

1. 水泵房水泵
   Water pumps in the pump room
2. 二段排水天沟。
   Drainage gutter in Block II
3. 伞壳餐厅屋面
   Roof of umbrella-shaped restaurant
4. 伞壳餐厅屋面雨水口的一铸铁圆形帽。
   A cast iron circular cap of the gutter of the umbrella-shaped restaurant roof

3. 给排水设计

泵房置于大会堂地下室独立的房间，供大会堂厨房、卫生间用水系统和消防水系统使用，地下室泵房设计有100立方米的钢筋混凝土水箱和给水泵、消防泵、室外喷泉泵，在钢屋架设置一个钢板水箱作为消防水箱。

一段屋面周边钢屋架采取放射形的，雨水从放射形屋架和梯形屋架之间做成的排水天沟，通过内落水管排出。且各段屋面均采用排水天沟。

但四段伞壳餐厅屋面为直接内排水，雨水口为一铸铁圆形帽。

3. Plumbing design

The pump room, set in a separate room in the assembly hall's basement, serves the kitchen, the toilet water system and fire water system in the assembly hall. In the pump room is a 100m^3 reinforced concrete water tank, a feed water pump, a fire pump, an outdoor fountain pump and a fire water tank made of a steel plate water tank set on the steel roof truss.

The surrounding steel roof truss in Block I has a radial shape. The rain water flows from the drainage gutter between the radial roof truss and the trapezoidal roof truss through the internal down pipe. The drainage gutter is on the roof.

But the roof of the umbrella-shaped roof restaurant in Block IV uses direct internal drainage, with a cast iron circular cap for the gutter.

2

3

4

项目施工
Project Construction

施工时的建筑师I 型塔吊
Architect type I tower crane during construction

一　施工综述

“班厦”整个工程的建筑、结构、水、暖、电，一直到同声传译系统的设备供应、安装调试等工作，全部由我国和斯里兰卡共同组织进行。我国总共派去了400余名技术工人，40余名工程技术和管理人员，仅木工就有60多人。而且各工种的工人都是五级工以上。当时的承建单位是北京市建工局第一建筑工程公司为主，因派不出那么多五级以上的工人，就从第五建筑工程公司、第六建筑工程公司抽调了一部分。

(I) Construction Overview

China and Sri Lanka jointly conducted work on the project, ranging from construction, structure, water, heating and electricity of the BMICH project and equipment supply, installation and debugging of simultaneous interpretation system. As all the workers had Grade 5 certification, China dispatched some 400 technicians and 40 engineers and managers, including over 60 carpenters. Originally, Beijing No. 1 Construction & Engineering Co., of Beijing Construction Engineering Group served as the major contractor, and because of the shortage of workers above Grade 5, Beijing No. 5 Construction & Engineering Co., and Beijing No. 6 Construction & Engineering Co., sent some workers as well.

当时我国施工机械化的水平不是很高，提供的施工机械有两台建筑师I型塔吊，6台400升混凝土搅拌机，还有十几台机动小翻斗，一台5吨的汽车吊，以及一些木工加工机械，钢筋加工机械，无齿锯和其他一些手推小车和手工工具等。所以大厦施工是以手工操作为主的。也可以说，“纪念班达拉奈克国际会议大厦”是世界上最大的手工艺品！当时斯方也提供了部分施工机械设备。

从设计到施工，项目组全体人员都是尽心尽力，兢兢业业，忘我工作。在施工上，提出了“绝对保证施工质量”的要求。

施工工期两年零五个月，期间无一例严重的施工事故。中国工人与斯里兰卡工人，在800多个日日夜夜里，冒着酷热，共同劳动，携手施工，发挥了极大的工作热情和积极性，高质量地按期完成了这项工程。

Due to the limited mechanization level in China at that time, only two type architect I tower cranes, six 400L concrete mixers, a dozen of powered mini dump trucks, one 5T truck crane, some wood-working machines, steel bar-processing machines, abrasive saws, trolleys and hand tools were available, along with some construction equipment provided by the Sri Lankan side. As a result, the construction of BMICH was largely accomplished by hand, which one could say Bandaranaike Memorial International Conference Hall is the largest handicraft of the world.

From design to construction, the project crew made all-out effort with conscientious care and dedication, thus giving an absolute guarantee in construction quality.

During the construction period of two years and five months (800 days), workers from both countries had to endure torturous heat and managed to avoid any serious construction accidents. They were both committed in their joint efforts to build and complete this project with the utmost passion and complete the building on time and with the highest quality.

1. “班厦”初建时情景
BMICH under initial construction

2. 大厦屋架吊装时
Hoisting the BMICH roof truss

3. 挑檐安装即将完工
Installation of the eaves is almost completed

二　施工经验

1. 钢结构安装

在屋面钢结构安装上，工程技术人员和工人们极大地发挥了他们的创造性。

大厦八角形的对角线是108米，一榀钢屋架要分三段或四段运到现场，而钢屋架该怎样安装，成为大厦专家组要解决的一个核心技术难题。因为，从塔吊到大厦中心线长达54米，远远大于塔吊的臂长，根本无法吊装。

(II) Construction Experiences

1. Installation of steel structure

When installing the steel roof structure, technicians and workers demonstrated their creativity.

Since the diagonal line of the BMICH octagon is 108m, a steel roof truss had to be transported to the site in three or four segments. However, hoisting was not allowed given that the distance from the tower crane to BMICH center line was up to 54m, far greater than the arm length of tower crane. Therefore, the problem of how to install the steel roof truss became a technical headache for the BMICH expert group.

1

2

3

4

经研讨，最后的办法是在标高8米的楼板上搭一个满堂的承重架子（因为做吊顶也需要承重架），在中间柱网边部再搭一个平台，用斯方提供的吊车把分段的钢屋架吊到这个平台上。然后用滚杠，让人推动屋架下垫的钢管一点儿一点儿向前移动，就这样让每段钢屋架就位、拼装，然后在与钢屋架垂直方向的两边平台各安装一台5吨慢转卷扬机开始工作。一边的卷扬机往上拽，另一边的往下松，因为不能把钢屋架拽过头，要保持平衡，确保钢屋架准确到位。还做了一个可移动的6米多高的拔杆用来吊装檩条。檩条吊装完成后，用斯方提供的大塔吊吊装向外放射的，像飞机的翅膀一样的外圈钢屋架。

2. 湿贴

“班厦”的40根24米高的八角形外廊柱子贴面石材，是中国山东出产的雪花白大理石。用雪花白大理石，还有把中国的雪花带到斯里兰卡，给那里带去凉爽的含意。在装饰外廊柱子时，用的是湿贴的施工方法，并要求柱子贴面外观接缝横平竖直，内无空鼓。因为有了空鼓，大理石容易脱落。为了保证没有一点空鼓，采用了分三次灌浆的方法。

After much discussion, the final solution was: firstly strike a packed bearing frame on the floor slab at 8m level (considering that the bearing frame was also required for the ceiling) and build a platform on the edge of the middle column grid to hoist the segmented steel roof truss to the platform with the crane provided by the Sri Lankan side; secondly, use a rolling bar to push the steel pipe under roof truss forward gradually to set each segment of the steel roof truss in place in order to assemble them; thirdly, install one 5T slow-acting hoist on both sides of the platform perpendicular to the steel roof truss, and pull one hoist while loosening the other to maintain the balance of and the exact position of the steel roof truss; fourthly, make a portable lever over 6m to hoist purlin; after that, hoist the airplane wing-like outer ring steel roof truss with the large tower crane provided by the Sri Lankan side.

2. Wet veneer overlaying process

Arabescato Corchia marble manufactured in Shandong Province China was used as the stone veneer for 40 octagonal exterior corridor columns as high as 24m in BMICH, to insinuate bring cooler temperatures to Sri Lanka as the name in Chinese is "snow white" marble. The wet veneer overlaying process was applied to decorate the exterior corridor columns and straight and even veneering exteriors were seamless to prevent hollowing (as marble can drop from hollowing). For this reason, grouting was applied three times.

1. 参加安装钢屋架的人们与负责安装的师傅及拔杆的合影
A photo of workers involved in steel truss installation, workers in charge of installation and lever

2. 准备安装钢屋架
Ready to install the steel roof truss

3. 平面梯形钢屋架形状，工人在开卷扬机
Plane trapezoidal steel roof truss shape, workers operating the hoist

4. 施工时在用斯方提供的大塔吊安装放射形钢屋架
Installing the radial steel roof truss during construction with the big tower crane supplied by the Sri Lankan side

5. 外廊柱子安装
Installation of exterior corridor column

6. 外廊柱落架的情景
Landing of exterior corridor column

3. 贴砖及木材

▪ 墙壁瓷砖

20世纪70年代，中国生产瓷砖的技术还不是很好，有些砖面不平，常有砖的四个角不在一个平面上的情况。为了保证施工质量，工人晚上就一块一块地挑选瓷砖，把质量好的、平整的瓷砖挑出来整块使用，有点残缺，不平整的瓷砖，也不扔掉，而是在边角处使用，正好把有缺陷的地方裁掉。这样，既保证了瓷砖墙表面平整，又为工程节约了很多材料。

▪ 水磨石地面

由于是八角形地面，斜角的部分很多，地面施工难度不小。在铺预制水磨石时，要把水磨石裁成斜角才能拼对出八角形。水磨石这种材料，由石子和混凝土压制而成，石子大小不一，且质地坚硬，切割时如操作不当，很可能有石子迸出，造成边缘不整齐，无法与其他的水磨石对缝。如果这样，这块预制水磨石就不能用了。为减少这种情况的发生，工人在划线时，尽量躲开有大石子的地方，在切割时，需要精心细致地推石材，减少齿锯对石子的冲击力，使石子尽可能地不迸出。这样既保证了切割后水磨石边的整齐，能正常使用，又减少了材料浪费。

3. Tiles and timber

▪ **Wall tiles**

In the 1970s, the technology for producing ceramic tiles was not very good, as the tile surface was uneven. To ensure construction quality, workers sorted out all the tiles one by one, separating the smooth, good quality ones to be used in whole from the irregular ones with defects for areas like corners instead of discarding them. In this way, a smooth tile surface was guaranteed and a considerable amount of materials were saved.

▪ **Terrazzo floor**

The existence of numerous oblique angles on the octagonal floor added to the difficulty constructing the floor, since octagonal shape could only be pieced together by cutting terrazzo in oblique angles when paving precast terrazzo, which is produced by pressing pebbles into concrete, with varied size and texture. If there were any problems with cutting the terrazzo, pebbles were likely to burst out and result in uneven edges, which would not make it suitable to place together with other terrazzo. If so, this precast terrazzo could no longer be used. To avoid this issue, workers firstly tried to avoid areas on the terrazzo with large pebbles when drawing lines, and then pushed the stone with great care during the cutting stage to reduce the impact force imposed on stone by the rack saw and to avoid pebbles from bursting out. With these efforts, an even cut terrazzo edge was guaranteed and there was less material wastage.

1. 一段三层男盥洗室
Men's room on 3F of Block 1

2. 四段伞壳餐厅的水磨石地面
Terrazzo floor of umbrella shell restaurant in Block IV

精细的马赛克地面
Exquisite mosaic floor

▪ 马赛克地面

在铺马赛克地面时，为了使施工水平达到顶级，工人更是精益求精。马赛克是一种很琐碎的装饰材料，粘贴面积小，很容易脱落。在使用过程中，要让马赛克的受力面一样，底层做的适度，才能保证将来不易脱落。当时，在铺马赛克前，要先在地面上喷水，再把干砂浆铺好，然后再浇水。浇水要恰当，太湿了不行，干了也不行，掌握这个尺度就是技术。然后将30厘米×30厘米一联的马赛克铺上，每一块都要用水平仪找平——铺水磨石地面时也是这样做的，再将纸联揭掉。为了使接缝笔直，工人还要用小改刀修改接缝。接缝全部做得很标准了，再用白水泥灌封。

▪ **Mosaic floor**

When paving the mosaic floor, workers were constantly perfecting their work to achieve the highest construction quality. As a small decoration material, mosaics can come off easily due to the small surface area for applying the adhesive. Therefore, it is crucial to have an even surface and a proper bottom layer during application. At the time, the practice was to spray water prior to paving, pave the mortar dry and then spray water again, neither too dry nor too wet. Then the mosaics were carefully positioned, requiring technique. After that, the 30cm×30cm mosaics were paved, leveled with a level gauge (the same practice as paving the terrazzo floor) and remove the paper. In addition, workers had to modify it with a small knife to enable a straight edge, and pour white cement to make all the joints seamless.

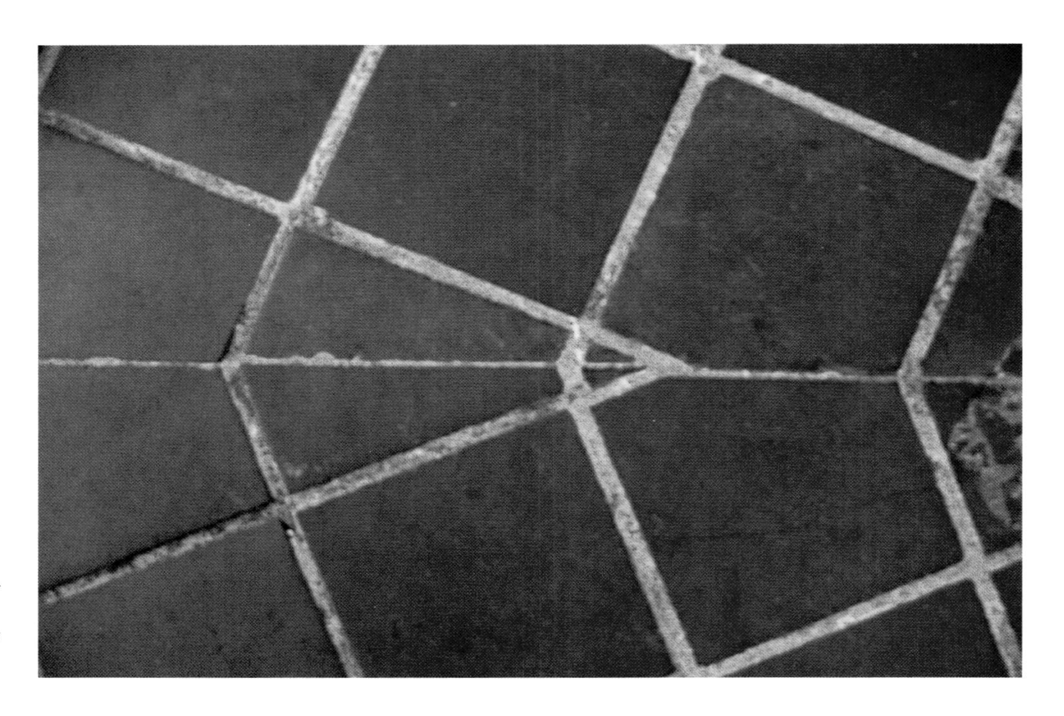

红缸砖地面的精细接缝
Exquisite joint of red clay tile floor

▪ 热带气候下的施工

在每一个施工现场，一定要考虑当地的自然条件和气候对施工的影响。斯里兰卡这个地方，气候炎热且潮湿，对施工有有利的一面，也有不利的一面。

有利的一面主要体现在打混凝土上。混凝土在高温的情况下凝固得比较快，达到强度的时间也会比较短。在国内的气候条件下，一般24小时之内为初凝，7天达到一个强度，14天达到一个强度，28天达到一个强度。斯里兰卡炎热且潮湿的气候使得拆模板的时间得以缩短，早一点拆掉模板就可以增加它的周转使用率，缩短施工时间且节省木材。

另外，浇灌混凝土时，由于温度较高，施工缝就要留得稍宽一点。做大台基室外地面铺红缸砖时，也要加宽砖缝来调节地面收缩的问题。

不利的一面来自木材方面。温度高，木材没有经过烘干处理，就很容易干裂。所以支完模板就要尽快绑钢筋，浇灌混凝土，否则就会有裂缝。做装修的木材，要用沥青浸透或长时间放置室外，让木材在自然气候中干燥，就不会开裂了。还有，当地的白蚁很厉害，对木装修危害很大。木材用沥青浸透后也能防治白蚁。

▪ **Construction in tropical climate**

For each construction site, the impact of local natural conditions and climate on construction must be taken into consideration. And for this project, the hot and humid climate of Sri Lanka presented advantages and disadvantages to the construction process.

The benefits of this kind of climate was in pouring concrete, which set faster and intensified in a shorter time period at a high temperature. The initial setting began within 24 hours and the concrete's strength was achieved in seven days, followed by 14 and 28 days respectively. While the hot and humid weather in Sri Lanka resulted in a shorter time in removing the form, this helped improve the turnover utilization rate, thus shortening the construction period and saved the use of wood.

In addition, with higher temperature, a wider construction joint was needed during the pouring of concrete. And, when paving the red clay tile of stylobate outdoor floor, the tile joints should be larger to adjust for floor shrinkage.

The disadvantages were found in the wood, which could easily break without proper drying treatment in high temperature. To avoid this problem, tie reinforcement to the wood and pour concrete immediately upon supporting it. As for wood for decoration, asphalt impregnation or long-term placement outside meant the wood needed to be dried naturally, in the hopes of avoiding cracks and also prevent the infestation by local termites via asphalt impregnation.

▪ 楼梯扶手

"班厦"的楼梯扶手拐角处，就都是用沥青浸过了的木材，而且是用整块木材手工雕刻出来的。

▪ 清漆磨退

"班厦"里面的墙壁大部分是木装修的。为了音响效果好，要在木条下边用矿棉填充，矿棉的上面要覆盖玻璃布，外面再加钢板网，把做得非常精细的梯形硬木条按等间距装在墙壁上。这些木条就位后全部做清漆磨退。

所谓"清漆磨退"就是在木条上刷一道清漆，等油漆干后，用砂纸把木头表面的漆打磨掉，然后再刷一遍油漆，再打磨掉……如此反复做7次，打磨砂纸的细度一遍比一遍细。最后，再用牙膏擦拭一遍，为的是将表面残留的细微毛糙之处磨光润色，起到油光锃亮的效果，完全是手工操作。完成后的效果，木头表面非常亮，亮得都可映照出影子来，用"油光锃亮"来形容都难以表达出这种亮的程度。

斯里兰卡的白蚁很多。为了防白蚁的侵咬，地板龙骨和地板以及屋面上木檩条的木材和木望板全部用沥青浸泡，预防白蚁。三十多年过去了，这些木材也很少被白蚁侵咬。

▪ **Stair railing**

The stair railing corner of BMICH used a whole piece of wood that was carved by hand upon asphalt impregnation.

▪ **Varnish rubbing**

Most of the walls inside BMICH are decorated with wood. To acquire ideal sound acoustics, mineral wools covered the wood with glass cloth first, and then metal mesh on the outside. Also, the exquisite trapezoidal hardwood strip had varnish rubbing on it when it was placed, and installed on the walls in equal intervals.

The so-called varnish rubbing is used to varnish the wood strip first, then polish the paint off the wood surface with sand paper (the fineness of which should increase after going over it several times). When the paint dries, the process is repeated seven times in total. Finally, toothpaste is used to polish the remaining minor coarse areas by hand to produce a shiny effect. In the end, the wood has such a shiny surface that can reflect a shadow.

To avoid the invasion of termites, which are common in Sri Lanka, floor keels, floors, and wood purlin for the roof and plank sheathing were immersed in asphalt. Now three decades past, and the wood is hardly damaged by termites.

1. 用整块木材手工雕刻出楼梯扶手拐角处
Hand carved whole piece of wood on the stair railing turn

2. "清漆磨退"梯形硬木条墙
Trapezoidal hardwood strip wall after rubbed with varnish

4. 遮阳板焊接

“班厦”的258根遮阳板是由钢筋混凝土制作而成的，每根由3段预制钢筋混凝土板组成的。置于8米标高楼板的梁端上。遮阳板每根高10米，面宽1.1米，前端厚度10厘米，后端厚度12厘米，呈梯形置放，间距1.1米。预制加工时就要做到表面非常光滑平整，棱角整齐，安装准确，垂直度好，要有很高的技术才能做出来。完工后，这774段遮阳板预制块，笔挺地围绕着大厦，令人赞叹预制钢筋混凝土能做到如此精细，简直如同手工艺品。这些遮阳板不但起到了遮阳的作用，还给大厦增添了美观、挺拔、高大的视觉效果。

5. 吊顶安装

标高8米的悬挑平台底下的吊顶，是用大块石棉板铺成后刷的白油漆，完成后看上去铺得非常整齐。但由于在吊顶的外边缘处，有一圈周边凹槽，内置日光灯管，试灯后发现，原本平整的石棉板吊顶因为灯的眩光照耀，显得凸凹不平了。为了避免产生眩光，又在灯的内侧加上一圈木条遮挡眩光。灯光又亮起来的时候，悬挑平台下的吊顶熠熠生辉，显得格外平整，如同一片舒展的云。

4. Welding of sunshade fins

The 258 sunshade fins were manufactured using reinforced concrete, and are comprised of three sections of precast reinforced concrete slabs and rest on beam ends of floor slabs with an elevation of 8m. Each sunshade board, 10m high, 1.1m wide with a front-end thickness of 10cm and rear-end thickness of 12cm, placed in a trapezoid with a spacing of 1.1m. During prefabrication, a smooth surface, neat edge, accurate installation and verticality were required, which was impossible without superior technique. Apart from the basic sunshade effect, it also created an elegant, upright and lofty visual impact for BMICH. Upon completion, BMICH is surrounded by 774 straight sunshade precast blocks, an admiration arises spontaneously for such exquisite treatment of precast reinforced concrete.

5. Ceiling Installation

The ceiling beneath the overhang platform with an elevation of 8m was paved with large pieces of asbestos boards and painted in white, showing a neat surface. But upon lighting the built-in fluorescent tube in peripheral grooves resting on the outer edge of ceiling, it was found that the originally neat asbestos board ceiling looked uneven because of the glare from the lighting. To this end, wood strips were added where the lights were to shield from the glare. Consequently, the ceiling gave the impression of a stretched cloud, exhibiting extraordinary flatness.

1. 工人在炎热的阳光下焊接遮阳板
   A worker is welding sunshade board in the scorching sunshine
2. 工人在安装遮阳板
   Workers install the sunshade board

1

2

遮阳板现状

Current condition of the sunshade

6. 防腐处理

因科伦坡地处沿海，空气潮湿，潮湿的海风对金属结构的腐蚀非常严重。一种是表面腐蚀，一种是穿孔腐蚀，都将造成严重后果。为预防这两种腐蚀，采用了金色的氧化铝。

施工中值得回忆的事例还有许多，从中我们不难看出，这座大厦的设计，完全符合了周总理对设计提出的三项原则。而且，建成这座大厦倾注着多少人的心血，是多少人用一双双灵巧的手精心建成的。而且，使用40年了，各个部位的结构和装修及各种设备，都还有着良好的应用效果。所以，说这座国际会议大厦是一件手工艺品一点也不为过，甚至可谓是一个世界上最大、最精致的手工艺品！

6. Anti-corrosion Treatment

Located along the coast, Colombo has high humidity in the air and the resulting corrosion (especially from surface and pitting corrosion) is adverse to metal structures. Golden aluminum was applied to prevent these two types of corrosion.

These cases are memorable. It is not hard for us to conclude that the design of BMICH is in strict compliance with the three principles raised by the late Premier Zhou Enlai, and that the completion of this building was a dream without the dedicated efforts of numerous workers. Now, four decades have passed, and the structure, decoration of each part and each piece of equipment are still in good condition. Thus, it is safe to say BMICH is a handicraft piece, the grandest and most exquisite handicraft in the world.

装修一览
Glimpse of Decoration

一　原貌集锦
(I) Collection of original appearance

1

2

3

1.纪念班达拉奈克国际会议大厦
Bandaranaike Memorial International Conference Hall

2. 班达拉奈克总理雕像
Bust of Prime Minister Bandaranaike

3. 礼仪大厅
Etiquette Hall

4. 主席台及代表席 会议大厅容纳1500人，大厅一层设有540个代表席，后部有100个列席代表席，大厅二层设有860个旁听席和记者席，厅内装有7种语言的同声传译。主席台：除会议设备外还有供演出戏剧、音乐用的相应设施
Rostrum and delegates' seats.The Assembly Hall can accommodate 1,500 people, with 540 delegates' seats and 100 for attending delegates at the back on the first level, 860 public seats and a press lobby on the second level, and simultaneous interpretation covering seven languages are supplied.Rostrum: along with conference equipment, as well as corresponding facilities for drama and musical performances

5. 大厅后部代表席及旁听、记者席。设备：代表席装有即席发言设施、扩音系统和供电台转播用的插播设备
Delegates' seats, public seats and the press lobby are located at the back of the hall. Equipment: delegates' seats are equipped with impromptu talk facilities, public address system and inter-cut equipment for radio broadcast

6. A（马蹄形）会议室，容纳438人
Committee room A with 438 seats

1

2

3

4

5

1. B会议室，容纳438人，坐席可灵活布置
Committee Room B with 438 seats and flexible seating layout

2. C、D会议室, 容纳50人
Committe rooms C and D, accommodating 50 persons each

3. E、F会议室,容纳30人
Committee rooms E and F, accommodating 30 persons each

4. 代表团休息厅一侧，代表团休息厅与上层旁听人员休息厅设有联系楼梯
One side of the Delegate Lounge, from which a staircase was built to go to the bystander lounge upstairs

5. 主席团休息厅，有收听大会议厅会议进行情况的设备
Presidium Lounge, equipped with devices listening the Assembly Hall

6. 旁听人员休息厅
Bystander Lounge

7. 记者休息室
Press Lounge

8. 休息室
Lounge

9. 休息廊与小卖部
Balcony Lounge and kiosk

10. 三段会议厅与会议室间的休息廊
Balcony Lounge between the Assembly Hall and Committee Rooms of Block III

11. 宴会大厅　有540个席位及扩声系统，供摄影报导的电源装置
Banquet Hall, equipped with 540 seats and public address system and power unit for photo journalists

1. 一段办公室之一
Office 1 in Block I

2. 一段办公室之二
Office 2 in Block I

3. 四段总理班夫人办公室
Office of Prime Minister Sirimavo Bandaranaike in Block IV

1. 四段总理班夫人的休息室
Lounge of Prime Minister Sirimavo Bandaranaike in Block IV

2. 四段总理班夫人休息室内中国双面绣屏风
Chinese double sided embroidery screen in the Lounge of Prime Minister Sirimavo Bandaranaike in Block IV

3. 四段总理班夫人休息室内卫生间
Washroom facilities in the Lounge of Prime Minister Sirimavo Bandaranaike in Block IV

1

2

3

4

5

1. 公用电话间
Telephone booth

2. 盥洗室
Washroom

3. 伞壳餐厅
Umbrella Shell Restaurant

4. 一段陈列厅入口
Entrance of the Exhibition Hall in Block I

5. 三段讲演厅兼电影厅
Lecture and Cinema hall in Block III

6. 电话机房
Telephone exchange room

7. 厨房
Kitchen

8. 大厦三段小会议室外观
Appearance of Committee Room in Block III which is next to Block IV

9. 四段办公楼内院
Administration building in Block IV

6

7

8

9

1. 二段立面，代表团办公楼
Façade of Block II, Delegation Office

2. 空调机房
AC plant room

3. 变电室
Transformer room

1. 变电室内
Interior of transformer room

2. 舞台调光室内
Interior of stage dimmer room

3. 大门入口传达室
Gate chamber at entrance

二　热爱“班厦”

(II) Love for BMICH

1. 竣工前，验收组人员在“班厦”前亲切留影。左起：李子武、由宝贤、王挺局长（左四）。总建筑师戴念慈（右一）
   A photo of acceptance group members in front of BMICH prior to completion. From left: Li Ziwu, You Baoxian, Director Wang Ting (fourth from left), chief architect Dai Nianci (far right)
2. 竣工前，由宝贤在“班厦”前留影
   A photo of You Baoxian in front of BMICH prior to completion
3. “班厦”后花园一景
   View of BMICH backyard
4. “班厦”一角
   Partial of BMICH
5. “班厦”天桥一角
   Corner of BMICH bridge

三　新貌集锦

2009～2012年期间，对“班厦”进行了全面的修缮,使得“班厦”又有了新貌。

(III) Collection of a new look

After a comprehensive refurbishment from 2009 to 2012, BMICH took on new look.

“班厦”立面新貌
New look of BMICH post refurbishment

1. 璀璨新“班厦”
Bright new BMICH

2. 水影夜映新“班厦”
New BMICH reflected in the water at night

1. 大会堂主入口新加的雨棚
   Newly added canopy at the main entrance of the Assembly Hall
2. 2012年5月4日在新“班厦”举办佛教节
   Buddhism Festival on May 4, 2012

1. 礼仪大厅新貌
New look of the Etiquette Hall

2. 三段通廊新貌
New look of the gallery in Block III

3. 大会堂楼座及顶灯新貌
Assembly floor seats and roof lighting

4. A（马蹄形）会议室新貌
New look of Committee Room A

1. 宴会厅新貌
New look of Banquet Hall

2. B会议室新貌
New look of Committee Room B

3. 休息大厅新风车灯
New windmill light fixtures in the Lounge

“班厦”效应
BMICH Effect

飞行器展览
Aircraft Show

“班厦”现在不仅是一个国际会议中心，同时也经常举行展览、演出、比赛、毕业典礼等各种活动。斯里兰卡人民对“班厦”的重视程度，从来宾隆重得如同过节的衣着打扮上就可见一斑：女士大部分穿着飘逸得体的各种颜色的民族服装——纱丽，再加上她们那婀娜的身姿，轻盈的步伐，为“班厦”增添了一道靓丽的风景线。绅士风度的男宾是西装革履；活泼可爱的学生是统一着装；佛界人士是袈裟披身；总之，来这里参加活动的各界人士，都是身着锦衣，兴高采烈、喜气洋洋。

从来来往往参加“班厦”举行各种活动的来宾、观众的言谈举止里不难看出斯里兰卡人民对“班厦”的向往和崇敬之情，就好像我国人民对人民大会堂的向往。

Currently, BMICH is not only an international conference center, but also a venue for holding exhibitions, performances, competitions and graduation ceremonies. The importance attached to BMICH by the Sri Lankan people is evidenced from the magnificent dress of the guests: ladies, mostly in elegant and national costume, the sari, and together with their graceful posture and light gait, contribute to the color of BMICH; gentlemen are in suits and ties, exerting gentleness; energetic students are in uniforms; Buddhists in robes. In short, people from all walks of life presented here are dressed, in high spirits and radiate with joy.

The yearning and admiration of the Sri Lankan people toward BMICH, just like the same emotions the Chinese people have towards the Great Hall of the People, can be glimpsed from the words and behavior of guests and the audience presenting activities sponsored by BMICH.

1. 在“班厦”举行佛教盛会
Buddhism Event in BMICH

2. “班厦”后花园的田园风光
Rural scenery of BMICH backyard

“班厦”院内的环境井井有条，修整得庄严、宏伟。可当你顺着一段小路绕到南侧，却自有一派曲径通幽、悠然自得的田园风光。静静的湖水、青青的芳草、高矮不一却又健壮挺拔的树木，还有一个全部用木材搭建的小桥，把人引向对岸别致的茅草别舍。好似远离了这繁华喧闹的城市文化中心，庄严肃静的政治文化中心，来到了一个幽静的农家村舍。在那里做一个深呼吸，你会感到那纯纯的空气是甜的，那微微的海风是润的，那淡淡的青草芳香还会没商量地扑到你的心里…… 沁人心脾，那里的一切都让人流连忘返。

BMICH presents an orderly, solemn and magnificent interior environment. While walking along a winding path to the south side, a secluded and carefree rural scenery is presented in front of you: a tranquil lake, lush grass, robust trees of all sizes and a bridge built of wood, leading people to the exquisite cottage on the opposite bank, far away from bustling urban cultural center and solemn political cultural core. Take a deep breath, and you may taste the sweet pure air, feel the soft touch of the sea breeze and smell the fragrance of the grass from the bottom of your heart. These senses are all refreshing and linger.

第三章
美丽的斯里兰卡
Chapter Three Beautiful Sri Lanka

美丽的斯里兰卡
Beautiful Sri Lanka

在美丽的斯里兰卡，有90%的人信奉佛教。信佛的人不杀生，宰杀牛羊都由天主教、回教的神职人员主持。因此斯里兰卡动物繁殖特别快，动物种类也特别多。人们也总是和动物有着亲密的感情，在建设“班厦”期间，人们在工作生活中，有着许多美好的印象，与动物之间也有着许多趣事。

In Sri Lanka, 90 percent of the population is Buddhist. Since Buddhists do not kill living things, Catholic and Islamic clergies oversee the slaughter of cattle and sheep. As a result, the animals there breed quickly and there are various species, and people always keep close relationships with animals. Evidence of work and life as well as numerous interesting episodes with animals during the construction of BMICH impressed people.

1. 美丽的斯里兰卡首都科伦坡
Colombo, the beautiful capital of Sri Lanka

2. 美丽的斯里兰卡印度洋海滨
Attractive Indian Ocean shore of Sri Lanka

生活
Life

由宝贤与不结果的西红柿合影
A photo of You Baoxian and his fruitless tomato plant

由于语言不同，工人怕走丢了，很少外出。但还要丰富工人们的业余生活，所以，经常在工地院子里搞一些轻松快乐的文化活动，解除工作疲劳，让人们忘掉炎热。

首先是每周六晚放电影，循环放映前面提到过的“三战一打击”。还有就是在大食堂里组织打乒乓球，木工们用施工废弃的木头做了个乒乓球台子，虽然条件很简陋，可打起乒乓球来，人们都是热火朝天的；还有中国象棋、扑克可供娱乐。

工人们还在食堂前种西红柿，为食堂增添些蔬菜。可不知为什么，这些西红柿只顾长高，足有一层楼那么高，像小树一样，却不结西红柿。

For fear of getting lost because of the language barrier, the construction workers rarely ventured out. But it was imperative to enrich the workers' leisure activities, and so relaxed cultural events were held in the site more often than not, hoping to relieve fatigue and have some respite from the scorching heat.

The events included: projecting movies every Saturday night (a circular projection of "Three Wars and One Fight" as mentioned above), playing table tennis in the dining hall on a surface made of discarded wood by the workers (despite the inferior conditions, everyone was very enthusiastic), playing Chinese chess and cards for entertainment.

In addition, workers planted tomatoes plants in front of the dining hall, hoping to supplement their diet with extra vegetables. But the tomato was fruitless, though it grew as high as one storey.

游玩
Tours

1972年的国庆节，在大厦工地工作的人们也放假三天。斯方为“班厦”的全体中国工作人员组织了一次旅游，去海底观鱼，参观野生动物园。

这次海底观鱼的景致留给人们的印象，要比想象的神奇美丽得多。海底观鱼时，四个人坐在一艘特制的船上出海，船边用铁皮制成，船底是玻璃。碧蓝的海水一眼能见底。大约8～9米深的海水底下，是纯洁无瑕的白珊瑚礁，美丽的热带鱼在清澈的海水中曼妙梭行。海底观鱼的美丽感受，实在让人难忘。

海底观鱼后，又去了野生动物园。观看了猴子等动物。最难忘的是人们看到了孔雀开屏的那令人欣喜的一幕。

游玩后，全体人员分两批乘坐斯里兰卡政府提供的大轿车，返回工地，又开始了紧张的施工建设。

During Chinese National Day in 1972, BMICH workers enjoyed a three-day holiday. The Sri Lankan side arranged a tour for all Chinese staff, to view fish in the seabed and visit Wildlife Park.

This fish-watching experience impressed people greatly, far beyond their imagination. Sitting in a specially made four-person boat (the boat was made of iron and the bottom made of glass), the workers appreciated the clear dark blue seawater. pure white coral reefs 8m-9m below the boat and beautiful tropical fishes gracefully swimming in the water. It was an unforgettable memory.

The following destination was Wildlife Park, where the workers were able to see several animals, like monkeys and luckily witnessed a peacock spreading its feathers.

Afterwards, they returned to the work site by sedans arranged by the Sri Lankan government in two groups and resumed construction work.

1. “十一”去游玩时，在蒂萨旅馆院内与斯方老局长的合影。左起：使馆翻译小崔，闫参赞，斯方工程部建设局副局长兼大厦斯方负责人，工程副组长兼总工程师由宝贤，工地翻译顾永清
A photo with the original director of Sri Lanka in the Tissa Hotel during Chinese National Day tour.From left: Mr. Cui (Embassy translator), Mr. Yan (counselor), deputy director of Construction Bureau of Engineering Department of Sri Lanka and BMICH principal of Sri Lanka,You Baoxian (deputy engineering supervisor and chief engineer), Gu Yongqing (site translator)

2. 海底观鱼
Viewing fish in the seabed

3. 由宝贤在野生动物园
A photo of You Baoxian taken in Wildlife Park

1

2

3

动物
Animals

在建设“班厦”期间，人们与动物之间有许多趣事、险事，让人们感到斯里兰卡更加美丽，那美丽至今都让人难忘。

有时出城办事，走到山麓一下车，猴子就跑过来，伸手向你要吃的。更有意思的是，钢屋架快要完工时，突然跑来一只猴子，跳到屋架中间，这位不速之客，玩乐了三天，才恋恋不舍地走了。

斯里兰卡的蛇很多，大使馆怕工人被咬伤，还专门从医务室派出一位医生，到有关部门去学了一周被毒蛇咬伤的急救技术，并带回当地特制的专治毒蛇咬伤的特效药。

还好，在“班厦”建设期间，虽然人与蛇多次相遇，但没有被咬伤过。只是有一次人蛇大战，好不热闹。事情是工人们洗澡时，可能是水汽熏蒸的作用，从顶棚上掉下一条蛇，吓得工人们乱作一团。还有一次，正是中午休息时间，突然，“吧嗒”一声从顶棚上“跳”下一条蛇来，搅了所有人的午休。

The interesting and adventurous episodes between people and animals during the BMICH construction period gave a deep impression of the beauty of Sri Lanka and impressed them immensely.

On some occasions, the workers went downtown on business and immediately got off the car at the foot of the mountain, where monkeys approached them and begged for food. More interestingly, when the steel roof truss was near completion, a monkey suddenly jumped onto the middle of the roof truss and lingered for three days.

There are numerous snakes in Sri Lanka, and the workers had a high chance of bites, so the embassy dispatched a doctor from the health center to the department concerned for a week to learn first aid for poisonous snake bites, and took back local miracle drugs for this ailment.

Fortunately, during the construction period, despite several encounters with snakes, no one was bitten. There were only a few encounters with snakes that were thrilling. A snake fell from the ceiling out of steam vent while workers were taking a bath, creating a chaotic scene. Another incident was a snake that "jumped" from the ceiling, disturbing the workers' lunch break.

1. 左为郝铺堂，右为由宝贤，身后背景是像山一样的白蚁窝
On the left is Hao Putang and on the right is You Baoxian standing in front of an anthill

2. 工程人员居住的工棚
The work shed inhabited by engineering staff

1～3. 斯里兰卡可爱的小猴子
A little monkey in Sri Lanka

4. 斯里兰卡小猴子图片欣赏：
左：正在树枝上打哈欠的小猴子
右：躺在树枝上正在玩耍的小猴子
Photos of little monkeys in Sri Lanka
Left: Yawning in the tree
Right: Playing while lying on the tree branch

斯里兰卡不但蛇多，蜥蜴也很多。在工人们住的房子前有一棵大树，在树干的树洞里，住着一条大蜥蜴。它经常探出头来往外看。

再有， 斯里兰卡的蚂蚁很多，白蚁窝堆得像小山一样，有一人多高。蚂蚁是咬人的，晚上睡觉时大家经常被蚂蚁咬醒。后来，当地斯里兰卡工人告诉中国工人，蚂蚁怕油墨味儿，大使馆给了我们很多报纸，算是解决了蚂蚁咬人的问题。

蚊子也很多，休息时，必须进蚊帐。那时，休息室也是办公室，办公时，脚下必须点上蚊香。否则，蚊子隔着裤子都咬人，有时还从裤口钻到裤子里咬。

工地里，工人们还养了18条狗。有条狗很有意思，专门捉老鼠吃。

Besides snakes, Sri Lanka is home to lizards. In front of the workers' house, there was a large tree which had a hole that was home to a giant lizard. It frequently exposed its head to look outside.

Moreover, there were plenty of ants there. The mountain-like anthill was nearly as tall as a man. Usually, the workers were woken up by ant bites. Later, this problem was solved when local workers informed them that ants were afraid of the smell of ink and the Embassy sent the workers lots of newspapers.

To keep annoying mosquitoes away, the workers had to hide in mosquito nets during breaks. During these times, the rest area was also used as the office, and so mosquito-repellent incense was a must. Otherwise, the mosquitoes would bite through fabric, even climbed into trousers.

The workers also raised 18 dogs on the site. Among them, one dog was very interesting because it lived on mice.

斯里兰卡全国各地都有象，自古以来，象深深地根植于斯里兰卡文化中。在斯里兰卡对大象还有这样的神话传说：相传很早时，有一个外国人，在斯里兰卡佛教庙堂里对佛不尊敬，寺庙里的大象就冲了上去，一脚将那外国人踩死了。

1975年斯里兰卡还建立了大象孤儿院，主要收养那些在野外失去母亲的幼象以及野象群失散的孤象。

Elephants are everywhere in Sri Lanka, and they are deeply rooted in the country's culture. A legend in Sri Lanka is that long, long ago, a foreigner offended Buddha in a Buddhist temple in Sri Lanka, and then an elephant immediately rushed forward and crushed the foreigner.

In 1975, an elephant orphanage, mainly adopting wild baby elephants that have lost their mothers and lonely elephants separated from their groups, was established in Sri Lanka.

1. 斯里兰卡大象
   Elephants in Sri Lanka
2. 斯里兰卡大榕树
   A big Banyan tree in Sri Lanka
3. 由宝贤在科伦坡庙宇里与大象的合影
   A photo of You Baoxian with an elephant in a temple in Colombo

第四章
重返“班厦”
Chapter IV Return to BMICH

重返“班厦”
Return to BMICH

当年“班厦”设计和施工负责人之一由宝贤先生与“班厦”管理处负责人班杜拉先生亲切握手
Mr. You Baoxian, one of the heads of BMICH design and construction, shakes hands with Mr.Bandhula Ekanayake, chief of BMICH

“班厦”要进行维修的消息，勾起许多人的那份难忘情怀，人们想要从不同的角度出发，记录“班厦”的故事。于是，2010年1月22日—1月27日，一些对“班厦”怀有特殊感情的人们，来到斯里兰卡，重返了阔别37年的“班厦”。

再次见到“班厦”，“班厦”还是那么的庄严，那么的富丽，那么的俊美！外廊柱子还是那么的高峻，那么的挺拔！那金色柱头，还是那么神圣地承托着宽大的挑檐……

Hearing that BMICH would be refurbished, a number of people recalled the unforgettable feelings they had and hoped to record the stories about BMICH from different perspectives. As a result, from January 22-27, 2010, people with special memories for BMICH came to Sri Lanka, returning to BMICH after a painful 37-year absence.

Coming to BMICH again, the place is still so solemn, splendid and beautiful! Just look at the tall, handsome and straight verandah column, the golden column cap supporting the broad eaves...

1. 由宝贤接受“班厦”管理负责人班杜拉先生授予的花环
Mr. You Baoxian receives a garland from Mr.Bandhula Ekanayake, chief of BMICH

2. 由宝贤与斯方官员亲切握手
Mr. You Baoxian shakes hands with Sri Lankan officials

3. “班厦”工作人员列队迎宾
BMICH staff line up to welcome the honored guest

4. 传统隆重的迎宾仪式
A traditional welcoming ceremony

斯里兰卡的阳光，依然是那样明媚地照耀着，斯里兰卡的海风，也依然那样轻轻地吹拂着。人们感受到的，是和当年一样的温暖，这温暖洋溢在人们激动的心中！

2010年1月25日14时，班厦管理处为“班厦”当年设计和施工负责人之一由宝贤的到来，举行了隆重的欢迎仪式。从门前台阶一直到门厅，都铺了红地毯，100多名“班厦”职员夹道欢迎，为由宝贤先生戴上美丽的花环和斯中友好徽章。那灿烂的花环在温暖阳光的照耀下，显得更加灿烂；那徽章至今仍在人们的心中熠熠生辉。

紧接着，迎宾队跳起欢快且最具斯里兰卡民族风格的迎宾舞。那锣鼓喧天、载歌载舞的热情欢迎场面真是让人心情激动，难以忘怀！随后，对“班厦”的各个大厅、会议室等处一一进行了参观。

The sun was shining brightly; the sea breeze was swaying softly; the warm feeling was just as before and rooted in people's hearts!

At 2pm on January 25, 2010, S.W.R.D. Bandaranaike National Memorial Foundation held a grand welcoming ceremony for Mr. You Baoxian, one of the heads of BMICH design and construction. With the red carpet rolled out from steps to the entrance hall and over 100 BMICH clerks lining the path, Mr. You Baoxian wore a beautiful garland and a China-Sri Lankan friendship badge. The garland seemed more brilliant than ever in the warm sunshine and the badge still shone in our hearts.

Then the cheerful national-style dance by the welcoming team was thrilling and impressive! Soon afterwards, the delegation visited all the halls and committee rooms in BMICH.

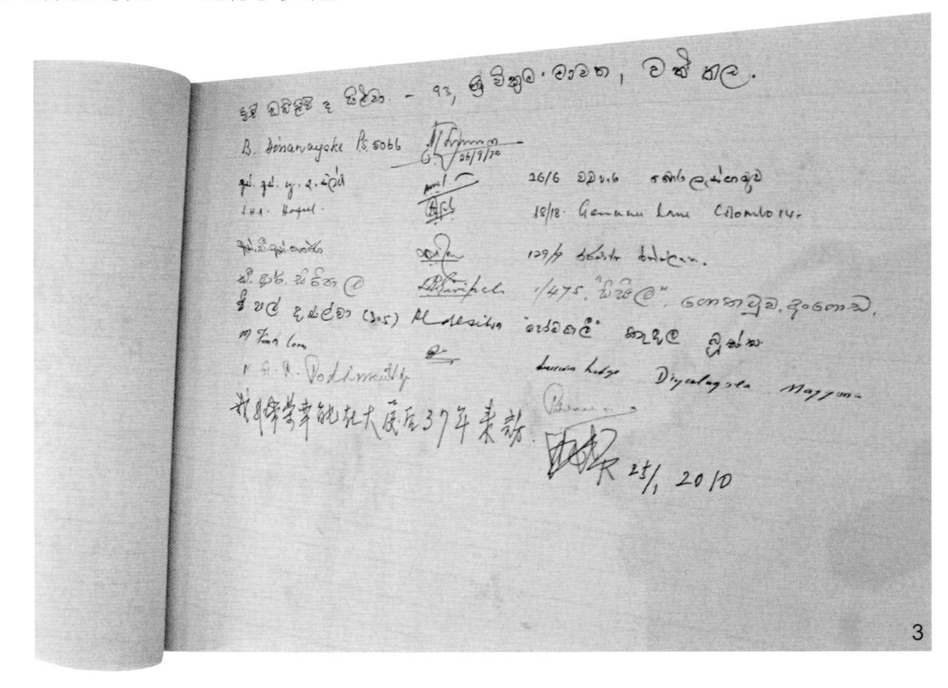

1. 由宝贤重返时在签名簿上留言左侧为由宝贤，右侧为“班厦”管理处负责人班杜拉
   Mr. You Baoxian signs his name in the visitors' book. On the left is Mr. You Baoxian and on the right is Mr. Bandhula Ekanayake, chief of BMICH
2. 欣赏“班厦”画册
   Looking through the BMICH picture album
3. 在签名簿上的留言
   Leaving a message in the visitors' book
4. 由宝贤与班杜拉在“班厦”大厅的合影
   A photo of Mr. You Baoxian and Mr. Bandhula Ekanayake
5. 由宝贤与张生、斯方负责人在“班厦”大厅的合影
   A photo of Mr. You Baoxian (center), Mr. Zhang Sheng (left) and a Sri Lankan principal (right) at the entrance hall of BMICH
6. 2011年2月26日“班厦”维修脚手架搭好时的壮观情景
   Stunning picture of BMICH with its maintenance scaffolding on February 26, 2011

愿“纪念班达拉奈克国际会议大厦”：
永远屹立在斯里兰卡的大地上！
永远屹立在印度洋那颗美丽的珍珠上！
永远屹立在中斯人民的心中！

May Bandaranaike Memorial International Conference Hall stand in Sri Lanka forever, on the beautiful pearl of the Indian Ocean and in people's hearts of both countries forever!

4

5

6

尾声：著者情怀
Epilogue: Author's Sensation

由宝贤1964年考察时，在美丽的斯里兰卡印度洋海滨

Mr. You Baoxian during his 1964 visit, sitting by the Indian Ocean in beautiful Sri Lanka

情深切

大海边，
美丽礁石吻浪喧。
吻浪喧，
印度洋水，
浩瀚无边。

斯里兰卡情深切，
40多载难忘却。
难忘却，
“班厦”情深，
泼洒诗绝。

2012.11.16晚

Deep Love

At the seaside,
Beautiful reefs kiss the ocean waves,
The waves are being kissed,
And the Indian Ocean waters,
Are vast and unbounded.

Our deep love for Sri Lanka,
Forty years cannot dilute it,
And it can never be diluted;
Is the deep love for BMICH in Sri Lanka,
Thus, a poem is written to express it.
On the evening of November 16, 2012

诗意：

情深切

我坐在大海边，聆听着美丽的礁石吻着海浪发出的悦耳的声音。那悦耳的声音啊！永远那么动听，那印度洋的水啊！浩瀚无边。

我对斯里兰卡的感情很深切，40多年了都难忘却，更难忘却，对“班厦”深深的、真切的热爱之情，在这里泼洒笔墨写诗一首，以表情怀。

2012.11.16晚

Poetic interpretation:

Deep love

Sitting by the sea, I listened to the euphoric sounds made by the crash of the ocean waves on the reef. The sound was pleasant to hear forever and the water of Indian Ocean was vast and boundless!

I have deep feelings towards Sri Lanka, which cannot be diluted over 40 years; and our profound and sincere love for BMICH could never be diluted. Therefore I wrote a poem to show my feelings.
On the evening of November 16, 2012

友谊颂

俊美班厦屹巍然，霞光披彩赞歌传。
金色柱头承天起，柱子正气定坤乾。
米胶协议破封锁，五项原则意志坚。
中斯人民携手建，象征友谊颂万年。

2013.1.3

Ode to Friendship

Pretty BMICH stands majestically, and the eulogy of colorful sunset is spread across the sky.
Golden column cap support the eaves, and the upright columns make things settled,
The Rice-Rubber Pact breaks the blockade, and Five Principles of Peaceful Coexistence is implemented firmly,
Chinese and Sri Lankan people work together, and the friendship is praised for its permanence.
January 3, 2013

卜算子 · 兄弟情

中斯友好情，
伟大跨长虹。
犹如兄弟情谊深，
赞美诗歌颂。

纪念班总理，
建厦哀思倾。
两国人民携手建，
友谊永象征。

2013.1.2.晚

Song of Divination: Brotherhood

The friendship between China and Sri Lanka,
is greater than ever,
It is comparable to brotherhood that
is praised forever.

In memory of Prime Minister Bandaranaike,
A hall is built as an expression of grief.
Two peoples work together,
A friendship is thus embodied forever.
On the evening of January 2, 2013

晚霞

夕阳红漫印度洋，谁言暮色惹感伤?
中斯人民共此景，喜迎明日新朝阳。

晚霞映照印度洋，寓意美好幸福长。
祝愿最美诗篇贺，中斯友谊万年长!

2013.1.4

Sunset Glow

The sunset permeates the Indian Ocean,
but it will not evoke sentimental feelings.
Two peoples appreciate the same scenery,
and they are joyful to welcome the new morning sun.
The sunset glow reflects the Indian Ocean,
and it means lifelong happiness.
Best wishes is expressed in poems,
so long live the China-Sri Lankan friendship!
January 4, 2013

永奏赞歌

中斯友谊践于行，班厦永奏赞歌声;
印度洋水深千尺，怎抵兄弟不老情。

2012.12.31

A Song Forever of Praise

The China-Sri Lankan friendship is put into practice,
so BMICH enjoys forever a song of praise;
The Indian Ocean is thousands of feet deep,
but it can not compete with unfading brotherhood.
December 31, 2012

以诗作画祝中斯友谊万古长青！

兰卡，

斯里兰卡，

我爱斯里兰卡，

那里海风习习，阳光明媚，

那里风光秀丽，那里景色让人陶醉！

她是镶嵌在浩瀚的印度洋上一颗美丽的珍珠！

晚霞，

美丽的晚霞，

美丽的晚霞柱子，

我爱美丽的晚霞柱子，

美丽的晚霞是班厦的灵魂和骄傲！

她寓意着美好的一天将要结束，更美好的明天即将来临！

美丽的晚霞啊！永远飘逸在班厦，永远永远飘逸在斯里兰卡！

金色，

尊贵的象征，

尽显尊贵的金色柱头，

它呈托着那宽大的挑檐，

它在斯里兰卡明媚的阳光下，熠熠生辉，光芒焕发！

雪花，

洁白的雪花，

40 根八角形外廊柱子，

它身躯纤细、笔直挺拔、举世无双！

它永远与班总理墓周围五根巍然屹立，象征和平共处五项原则的柱子遥相呼应。

中国雪花白大理石把它装饰，寓意把中国洁白的雪花带到斯里兰卡！

班厦，

我爱班厦，

我爱俊美的班厦，

它庄严、富丽，它俊美、高雅！

中国无偿援建的“纪念班达拉奈克国际会议大厦”！

她永远是中斯两国人民友谊的象征！她永远是中斯两国人民心中的明珠！

《中斯友谊的象征》

“纪念

班达拉

奈克国

际会议

大厦”

援建工

程技术

及纪实

Wish an everlasting Sino-Sri Lankan friendship in the form of a poem!

Lanka,
Sri Lanka,
I love Sri Lanka,
Where the sea breeze blows, the sun is shining,
Where beautiful hills and waters unfold, the scenery is intoxicating!
Undoubtedly, she is a beautiful pearl set in the vast and boundless Indian Ocean!
The sunset glow,
The beautiful sunset glow,
The beautiful sunset glows columns,
I love the beautiful sunset glowing columns,
The beautiful sunset glow is the soul and pride of BMICH!
It signifies the end of a beautiful day, and the approach of a better tomorrow!
Oh, the beautiful sunset glow, may it flow through BMICH forever, and in Sri Lanka forever and ever!
Golden color,
The symbol of nobility,
Represents the distinguished golden column,
That which supports the broad eaves,
And is shining and refreshing under the sunshine of Sri Lanka!
Snow,
Is pure white,
The 40 octagonal colonnades,
Are slender, straight and unique in the world!
They stand in front of the tomb of late Prime Minister Bandaranaike forever, symbolizing the Five Principles of Peaceful Coexistence!
They are decorated with snow-white marble, bringing white snowflakes to Sri Lanka!
BMICH,
I love BMICH,
I love beautiful BMICH,
Which is solemn, splendid and elegant!
Bandaranaike Memorial International Conference Hall is a gift from China,
A permanent symbol of China-Sri Lankan friendship, a bright pearl in the hearts of both countries!

张生，中国山西国际经济技术合作公司总工程师，"班厦"维修改造项目技术组组长；与"班厦"负责人班杜拉先生（左）和总工（右）的合影。

Mr. Sheng Zhang, is the Chief Engineer of China Shanxi International Economic & Technical Cooperation Corp., and the Director of the "Ban Sha" (BMICH) maintenance and renovation project, in the picture at left he was with the "Ban Sha" (BMICH) leader Mr. Bandhula Ekanayake (on left) and the Chief Engineer (on right).

段明宽，"班厦"维修改造项目总工程师。在"班厦"维修期间和斯里兰卡小朋友在"班厦"。

Mr. Mingkuan Duan, the Chief Engineer for "Ban Sha" (BMICH) maintenance and renovation project, is companying with Sri Lankan children at "Ban Xia" (BMICH) during the project period.

王丽丽，现在北京宝兰厦建筑工程技术咨询中心及中外建工程设计与顾问有限公司八所办公室任职，曾在参四任职。

Ms. Lili Wang, is now the office staff of the 8th department of China International Engineering Design & Consultant Co., Ltd, and Beijing Baolansha Engineering Tech Consulting Center. She had served in Cansi.

汪斌，2001年同由宝贤先生合作，成立广州第一家甲级民营建筑设计事务所，现为广州宝贤华瀚建筑设计有限公司，任副董事长、总经理。

Mr. Bin Wang, collaborated with Mr. Baoxian You, to establish the first Class A private architectural design firm in Guangzhou in 2001, Guangzhou BaoxianHuahan Architectural Design Firm, and became the Vice Chairman of the board and the General Manager.

杨建新，专业：工民建。高级工程师，国家二级注册建造师。曾多次被评先进工作者。曾任南通三建二分公司副总经理。现任江苏盛大建设工程有限公司董事长（法定代表人）。

Mr. Jianxin Yang, majored in Civil Engineering, Senior Engineer and Level Two registered builder. He has been awarded multiple times as Model Worker. He used to be the Deputy General Manager of the 2nd branch of Nantong Sanjian. He is now the Chairman of the board of Jiangsu Shengda Construction Engineering Co., Ltd.

刘俊，现任广州宝贤华瀚建筑工程设计有限公司工作南昌分公司总经理，并担任南昌市建筑艺术委员会委员，江西省重点工程办公室资深专家。“九江市中医医院南院”项目获2011年江西省勘察设计优秀奖和江西省首届十佳建筑殊荣。

Ms. Jun Liu, is now the General Manager of the Nanchang branch of Guangzhou Baoxianhuahan Architectural Engineering Design Inc. She is also a member of Nanchang Architectural Arts Committee, and a senior expert of the key projects office of Jiangxi Province. She is honored the Excellence in Survey and Design and Top 10 buildings for the project of "Jiujiang Traditional Chinese Medicine Hospital South Branch" in Jiangxi Province, 2011.

敖洪，1993年7月毕业于上海同济大学建筑系，拥有国家一级注册建筑师资格。现任广州宝贤华瀚建筑工程设计有限公司设计总监。主要业绩：富力公园28、公路大厦等。

Mr. Hong Ao, graduated from Department of Architecture, Shanghai Tongji University in July 1993, has obtained the National Class A registered architect. He is currently the Design Director at Guangzhou Baoxianhuahan Architectural Engineering Design Inc. His main projects includes the Fuli Park 28, Gonglu Building etc.

丁宁，广州宝贤华瀚建筑工程设计有限公司执行董事，总建筑师，一级注册建筑师。擅长城市设计、区域规划设计以及大型公共建筑设计。与中国优秀地产企业保利、富力、中海等保持长期良好的合作关系。

Mr. Ning Ding, Executive director, Chief Architect and Class A registered architect at Guangzhou Baoxianhuahan Architectural Engineering Design Inc., is expertised in urban design, regional planning, and large scale public facilities design. He maintains good long-term collaborations with Chinese excellent real estate developers, such as Baoli, Fuli, Zhonghai etc.

刘畅，2008年本科毕业于清华大学，现于耶鲁大学攻读博士学位。

Mr. Chang Liu, graduated from Tsinghua University in 2008 with a B.S., is now studying at Yale University for his Ph.D. degree.

付峥，2013年毕业于东北师范大学物理学院，获得学士学位。

Mr. Zheng Fu, graduated from School of Physics, Northeast Normal University in 2013 with a B.S..

《中斯友谊的象征》编委会

Editorial Board of A Symbol of China-Sri Lankan Friendship

主　　编：由宝贤

副 主 编：王丽丽　党　政

编　　委：张　生　张仲良　孟建国　张大明　杨建新　汪　斌　敖　洪　刘　俊　丁　宁

特邀编委：班杜拉・埃克纳亚克

作　　者：由宝贤　党　政　张　生　段明宽　王丽丽　刘　畅　付　峥　杨建新　汪　斌　刘　俊　敖　洪　丁　宁

翻　　译：喻蓉霞　党　政　刘　畅　付　峥

模型双语图：关海燕　詹顺顺

协助单位：中国建筑设计研究院
中外建工程设计与顾问有限公司
华森建筑工程设计顾问有限公司
北京筑邦建筑装饰工程有限公司
广州宝贤华瀚建筑工程设计有限公司及南昌分公司
江苏南通三建集团有限公司
北京宝兰厦建筑工程技术咨询中心

特别鸣谢：中国驻斯里兰卡大使馆

EDITOR-IN-CHIEF: Baoxian You

DEPUTY EDITORS: Lili Wang, Zheng Dang

EDITORIAL BOARD: Sheng Zhang, Zhongliang Zhang, Jianguo Meng, Daming Zhang, Jianxin Yang, Bin Wang, Hong Ao, Jun Liu, Ning Ding

GUEST EDITOR: Bandhula Ekanayake

AUTHORS: Baoxian You, Zheng Dang, Sheng Zhang, Mingkuan Duan, Lili Wang, Chang Liu, Zheng Fu, Jianxin Yang, Bin Wang, Jun Liu, Hong Ao, Ning Ding

TRANSLATORS: Rongxia Yu, Zheng Dang, Chang Liu, Zheng Fu

MODEL BILINGUAL MAP: Haiyan Guan, Shunshun Zhan

ASSISTING UNITS: China Architecture Design & Research Group
China Foreign Architecture Engineering Design and Consulting, Co., Ltd.
Huasen Architecture Engineering Design Consulting, Co., Ltd.
Beijing Zhubang Architecture Decoration Engineering Co., Ltd.
Guangzhou Baoxian Huahan Architecture Engineering Design, Co., Ltd., and the Nanchang Branch.
Jiangsu Nantong No.3 Architecture Group, Co., Ltd.
Beijing Baolanxia Architecture Engineering Technology Consulting Center

SPECIAL ACKNOWLEDGEMENT: Chinese Embassy in Sri Lanka

■ 感谢为本书提供帮助的人们！

Sincere gratitude to those who give strong supports for publishing this book!